The Digital Agenda

By

Lionel Eddy

Table of content

Introduction

In recent years, the rapid growth of digital technology has transformed the way we live and work, with the digital economy becoming an increasingly important driver of economic growth and innovation around the world. Governments are therefore rapidly realizing the necessity of creating policies and initiatives that support the expansion and advancement of the digital economy.

One of the key areas of focus for governments in this regard is the development of cashless societies, social credit systems, digital ID, and currency. These policies are aimed at streamlining transactions, enhancing security, improving accountability, and promoting economic growth.

Cashless societies, for instance, can help reduce the costs associated with handling cash, such as security and transportation, while also making transactions faster and more convenient. Digital currencies, on the other hand, have the potential to reduce the cost of cross-border transactions and increase financial inclusion by providing access to financial services to underserved populations.

Social credit systems, which use big data and artificial intelligence to assess an individual's creditworthiness based on a range of factors such as their financial history, behavior, and social interactions, can help promote accountability and transparency in financial transactions.

However, these policies also raise important ethical and privacy concerns, such as the potential for discrimination and infringement on individual rights. Therefore, it is crucial for governments to carefully consider the potential benefits and drawbacks of these policies and to develop comprehensive regulatory frameworks that balance the need for innovation with the need to protect individual rights and freedoms.

In this context, the role of government policies is essential in shaping the future of digital society. Governments have a critical role to play in fostering innovation, promoting economic growth, and protecting the rights and interests of citizens. By developing policies and programmes that strike the right balance between these objectives, governments can help to ensure that the digital society evolves in a way that is sustainable, equitable, and inclusive.

The book will provide readers with a better knowledge of the many digital ID systems that are being created and put into use throughout the world, as well as their potential advantages and difficulties, as well as the problems with privacy, security, and interoperability. The many kinds of digital currencies, such as cryptocurrencies, stable coins, and central bank digital currencies, as well as their potential advantages and disadvantages, such as concerns about regulation, security, and financial stability, will also be covered for the readers. The policies and impacts of major digital firms like, Apple, Facebook, Google, and Microsoft, as well as issues with data privacy, competition, and the social effects of technology, will also become clearer to readers. Throughout this book, readers will obtain a thorough grasp of the different initiatives and technologies that are influencing the development of a digital society, as well as the moral and privacy issues raised by their use.

A brief overview of the various programmes and technologies that will be explored in the book

The development of the digital economy has given rise to a range of new technologies and programmes that have the potential to transform the way we conduct transactions, interact with each other, and engage with the world around us. In this book, we will explore several key programmes and technologies that are shaping the future of the digital society, including;

AI: AI or Artificial Intelligence is the ability of a digital computer or computer-controlled robot to perform tasks normally associated with intelligent beings. The term is often used for projects that develop systems that characterize human intellectual processes, such as the ability to reason, discover meaning, generalize, or learn from past experience. AI can have an impact on workers ranges from applications that improve worker safety and health to potential hazards that need to be managed. We will look into the impact of AI in our society as well as the potential risks regarding the advancement of AI in the digital age.

A Cashless Society: A cashless society is one in which all transactions are conducted digitally, without the need for physical cash. This can include mobile payments, digital wallets, and other forms of electronic payment. We will examine the benefits and drawbacks of cashless societies, including the potential for greater efficiency, security, and convenience, as well as concerns related to privacy, cybersecurity, and financial exclusion.

Social Credit System: A social credit system is a mechanism that uses big data and artificial intelligence to assess an individual's creditworthiness based on a range of factors such as their financial history, behavior, and social interactions. We will review the various types of social credit systems in use around the world, analyze their potential benefits and drawbacks, and explore the ethical and privacy concerns associated with their implementation.

Digital Authoritarianism: Digital authoritarianism can be loosely defined as the use of digital information technologies by states for social control, oppression, surveillance, and otherwise strengthening their rule. Digital technologies have greatly expanded the toolbox of social control. Data collection systems, advanced biometric technologies, and advanced artificial intelligence data processing systems enable accurate and comprehensive citizen tracking and analysis through massive collection, analysis, and classification of data, enabling the government to achieve the granularity and scope of monitoring efforts.We will look into how this can have ramifications for our freedoms and our rights.

Digital ID: A digital ID is a secure and reliable way to verify an individual's identity in digital transactions. We will examine the various types of digital ID systems that are being developed and implemented around the world, including biometric identification, blockchain-based systems, and mobile-based systems. We will also explore the potential benefits and challenges associated with these systems, including issues related to privacy, security, and interoperability.

Digital Currency: Digital currency is a form of currency that exists entirely in digital form, with transactions conducted through a decentralized digital ledger known as a blockchain. We will review the various types of digital currencies, including cryptocurrencies, stablecoins, and central bank digital currencies. We will also explore the potential benefits and drawbacks of digital currencies, including issues related to regulation, security, and financial stability.

These companies have significant resources and control over vast amounts of data, which they use to develop new products and services as well as to generate revenue through targeted advertising. However, their dominance has also raised concerns about issues such as data privacy, competition, and the impact of technology on society.

As big tech companies continue to shape the tech landscape, there is increasing scrutiny around their practices and calls for greater regulation to ensure they operate in the best interests of users and society as a whole.

Transhumanism: Transhumanism is a movement that advocates the use of technology and science to enhance human beings' abilities beyond their natural limits. This includes technologies such as genetic engineering, nanotechnology and artificial intelligence. The goal of transhumanism is to improve human health, intelligence, and longevity, and to create a post human society in which individuals have greater control over their own biology and evolution. Transhumanists believe that the limitations of the human mind and body can be overcome through the use of technology, which can lead to a better future for humanity.

However, there are also concerns about the ethical and social implications of transhumanism, including the potential for deepening inequality and the loss of what it means to be human. Critics argue that the pursuit of transhumanism could lead to a dystopian future in which these enhancements are only available to the rich and powerful, leading to a growing divide between rich and poor.

Overall, transhumanism is a complex and controversial movement that raises important questions about the role of technology in shaping the future of humanity. Readers also learn about the transhumanist movement and how science and technology are being used to expand human potential beyond current possibilities. Participants will learn about the ethical and social implications of this movement.

Chapter 1

AI and the Fourth Revolution

AI has significant effects on democracy, and the right of individuals to a personal life and dignified existence. For instance, if AI were employed to determine the political ideologies of individuals, then our society would have might be manipulated. Political strategists could utilize this data to recognize which voters are most likely to be convinced to switch allegiance, or to increase or decrease their support the probability of them actually voting, and then spending resources on trying to convince them to do so. This sort of strategy has been said to have had a significant impact on the results of recent elections in the UK. and the US Alternatively, if AI can differentiate the emotional states of people and determine when they are lying, these individuals would have a more difficult time could be victimized by those who don't share their faith, or who bully others because of failed attempts at career advancement. In some cultures, it may lead to incarceration or death.

The death rate at the hands of the state has had a significant impact on the results of recent UK elections and the US The capacity for recognizing faces is said to be employed in China not only to recognize individuals, but also to identify them but to understand their moods and how they focus in both reeducating camps and regular schools. This technology would allow students to be punished for not paying attention or for disregarding the law prisoners that appear to be unhappy with the reeducation process. Unfortunately, governmental agencies are not necessarily dedicated to serving their citizens. The Chinese government has already employed tracking devices that would place over a million of its citizens in camps for the purpose of reeducating them for the criminal behavior expressing their Muslim faith There is a possibility that governments concerned with diversity will utilize AI to silence, restrain and negatively impact individuals. The right to speech and expression is considered a fundamental component of democracy.

This could be useful profoundly affected by AI

AI is touted by technology companies as a solution to problems issues like hate speech, violent behavior and digital deceit, the analysis of speech online is increasing in prevalence, and is often used to gauge the nature and tone of speech professionally trained to automate content removal. Governments has also expressed an interest. In employing AI to differentiate fake news and improve the country's reputation on social media this is a concerning trend, given the limited capacity of machine learning to understand the tone and context of speech. Automated that content removal carries the potential for speech censorship that is legitimate; this danger is increased by the fact that it is undertaken by private corporations, sometimes in response to government instruction. Heavy monitoring has a negative effect on the freedom of expression, this is because it promotes self-censorship. AI is derived from humans, which means it is susceptible to error. Systematic favoritism may be present as a result of the result of the data employed to train systems, or the value of the system developer or user. It is typically triggered when machine-learning apps are built on data that only shows certain demographic classes, or which represent societal preferences.

Several instances have been documented attention to promoting accidental social biases, which have been then followed or repeated automatically augmented by AI technology. The concentration of technological, economic and political power in a few large corporations would grant them an unfair influence over government decisions. However, the implementation and adoption of AI would lead to a different outcome. Other dangers to democracy include the potential loss of democracy through violence, vandalism, or other means. Throughout history, political parties have advocated for the office have relied on limited, anecdotal evidence and surveys to understand what they were saying. The electorate is considering. Now that Big Data is available, political figures have the opportunity to access enormous datasets information that allows them to focus on specific groups of voters and create messages that are intended with them the most.

People have been worried about the loss of jobs to technology for centuries. Automation, followed by mechanization, computers, and more recently AI and robots have all been involved expected to destroy jobs and lead to long-term damage to the job market. Automation is a wide range of technologies that reduce human intervention in processes, it was previously been employed to replace human labour in the short term instead automation has led to the creation of long-term jobs.

However, there is universal agreement that artificial intelligence and other associated technologies would lead to mass unemployment during the following generation two centuries. As computers become more creative, and intelligent, versatile, more jobs will be dependent on technology and more positions will be obsolete. The technology of artificial intelligence is growing in popularity around the world.

Additions continue to emerge, from the start of deepfake videos that diffs the boundary between the real and the fake, to cutting-edge computers that can surmount the greatest poker players on the planet in multiplayer games. Businesses utilize AI to enhance the processing of analytics in cities; officials in cities utilize AI to observe traffic congestion and oversee the efficient measurement of smart energy.

However, increasing numbers of states are utilizing advanced AI technology that facilitates surveillance, tracking, and survey of citizens in order to achieve a variety of goals that are lawful, but also goals that violate human rights, and there are many of them that are ambiguous ground. In order to adequately discuss the effects of this technology, it is essential to understand it first. The location of these instruments and how they are being employed. Unfortunately, this information is limited. Chinese products are often sold with soft loans that encourage government purchase their tools. These strategies are particularly pertinent to countries like Kenya, Laos, Mongolia, and others. Uganda and Uzbekistan would have otherwise lacked the opportunity to utilize this technology.

This poses a troubling question queries regarding the degree to which the Chinese government is funding the purchase of land advanced technology for repression. Governments in countries with absolute power are more powerful. More susceptible to abuse than government surveillance in liberal societies. Some dictatorial for example, China, Russia, and Saudi Arabia are utilizing AI technology to mass-produce to observe and record. Other countries that have a poor human rights record have relied on AI to accomplish their goals monitoring and evaluating the behavior of criminals in more limited ways that promote repression.

However, all political situations are susceptible to illegally using AI technology to surveil the political environment. Liberal democratic societies are actively utilizing AI to regulate borders and capture potential criminals, observe citizens for criminal behavior, and expose suspected terrorists to crowds. This doesn't necessarily mean that democracy is utilizing this technology in an illegal manner.

Is there a robust legal tradition of law and an independent system of justice?

That would provide a means of security for citizens of democratic countries. However, advanced democracies are attempting to balance the interests of security with the protection of civil rights. In America, more cities are implementing advanced monitoring systems.

Fourth Industrial Revolution

The world is close to a new technological revolution that will have a significant impact on our way of being, conducting business, and communicating with others. In its magnitude, scope and complexity, the transformation/ changes will be different than anything humans have experienced thus far. Still, no one has a clear idea of how it will develop, but one thing is apparent: our response should be universal and comprehensive, including all participants in international politics, from the public to the private sector to the intellectual and civil society. Klaus Schwab, the CEO and Chairman of the World Economic Forum in Davos, describes these technological advances as the fourth industrial revolution.

During the first industrial revolution, water and steam power were employed to replace manual labor with machine labor, this led to the development of industrial production. During the second revolution, electricity was employed to expand and enhance the volume of mass production, this led to the development of electronics and information technology. These technologies were then employed to automate production. Right now, the third industrial revolution is being supplanted by the fourth, a digital one that has evolved and grown throughout the middle of the last century. This has been primarily characterized by the blurring of traditional distinctions between the material, digital and biological worlds.

Since 1784, steam power, water, and mechanical devices for manufacturing have been the first to revolutionize industrial production. Since the time of the 70's, the division of labor, electricity, and mass production has had a significant effect on the second. Since 1969, electronics, information technology, and automated production have become the most common features of the third generation. Is it possible to determine the primary and essential characteristics of the following revolution? Three markers are present that indicate that the present alterations do not follow the third industrial revolution, but rather represent the beginning of the fourth. These are the speed, magnitude of alterations, and their overall effects on the entire system. The speed at which new discoveries occur and technological innovations are made has never been documented before. Contrasting with the previous revolutions, the rate of development of the fourth revolution is greater than the previous ones. Additionally, almost all of the industrial sectors in every country are being updated, and the magnitude and scope of the changes are indicative of a transformation of the entire production, management and administration systems. The potential for thousands of millionaires, connected via cell phones with cutting-edge computing capabilities, to have access to knowledge and to store information is limitless. These opportunities will be enhanced by the discovery of new technologies in multiple areas, including artificial intelligence, robots, autonomous transportation, the Internet of Things, 3D printing, Nano-biotechnology, materials science, energy technology and quantum computing.

Challenges and Opportunities

The fourth industrial revolution is similar to the previous three revolutions in that it can increase the global income and improve the quality of life of humans on Earth. Consumers with material resources and access to the digital world are the ones who already enjoy the benefits of its technologies, products and services. Through these resources, new products and services are created that increase efficiency and comfort for everyone. Taxi requests, flight ticket bookings, marketing, payment processing, music, film and computer games, all of which are accessible from a remote location. In the future, new technological innovations will lead to significant changes in the way industrial purchases are conducted, with a long-term increase in efficiency and productivity.

Transport and communication expenses will be decreased, logistics and global supply chains will be enhanced, trade margins will be decreased, all of which will lead to the expansion of new markets and economic growth. Additionally, the revolution may lead to an increase in inequality due to alterations in the labor market. Either automation, which substitutes for manual labor, or direct employment of a worker for a machine, can increase the disparity between the income of capital and the income of employees. However, this potential replacement of employees may in the future lead to an increase in safe and beneficial jobs and occupations. This shift has never been completed smoothly with the assistance of social expenditures that are significant. Additionally, most of them occurred in countries that failed to make the transition too early. As a result, the first industrial revolution led to deindustrialization in India there was a significant lack of technology in Russia and Turkey.

The second industrial revolution triggered the Great Depression, which led to significant social damage not only in the US, Germany, Argentina, but also to mass starvation in the colonies. We have not yet documented the full range of effects of the third industrial revolution. However, the most rudimentary estimates indicate that deindustrialization of the post-Soviet space, a series of problems in Latin America are the results of this shift, in which these countries have been behind. Other examples include China and Korea, which employed technology through effective policy towards industry and managed to become world leaders after the third technological revolution.

The earliest estimates indicate that at least 20% of the jobs have been eliminated. Globalization has led to very unequal job losses in different nations. The practice of over production during the transition from the global crisis did not allow people to take part in the released labor force on a global scale through increased production. Additionally, the average annual economic growth of 3 to 4 percent does not address any of the issues. To this point, it's hard to think about the probability of developing one scenario or the other, but there are historical prerequisites that suggest that the end result will be a mixture of both versions.

However, it's obvious that talent is more important to the production function than capital. This will lead to the division of the labor market into two separate segments: "low-skilled labor/low salary" and " high-skilled labor/high salary". As a result, social tension will increase. Other than its significant economic contribution, disparity will have a significant social component in the context of the fourth industrial revolution. The greatest benefactor is the person who provides intellectual and financial capital investors and savers these individuals explain the increasing gap between people who require capital and labor. As a result, technology is one of the primary factors of stagnation, along with decreasing income, this is also true of countries with high income: the demand for highly skilled workers is increasing, but it is diminishing for medium skilled workers. As such, the job market will have a high demand for skilled labor, but little demand for unskilled labor. The middle class will have no demand for them.

Also, digital technology's spread and the dynamics of social media in the dissemination of information could augment resentment. Today, over 30% of the world's population relies on social media platforms for communication, education and information distribution. In a perfect world, these relationships would facilitate the construction of bridges between different cultures and, as a result, lead to unity and consistency. However, these associations also could promote unrealistic or utopic views of the personal or collective success, as well as suggest and communicate extreme (extremist) concepts and ideas.

Today, significant problems still exist in the concept or predictions of how fast innovation will progress, and the increase in speed of change, both of which are considered paramount in the planning of any organization. These issues are the source of constant confusion for the most skilled and informed participants. All things considered, it's clearly demonstrated in every sector of the economy that the technology that marked the beginning of the fourth industrial revolution has a significant impact on business. The fourth industrial revolution is already underway, it is now expanding and the development of robots will further automate the economy and increase the use of previously-unexposed technologies, respectively. In the supply and logistics of many industries, we can observe the introduction of new technology, this technology enables a completely new approach to serving existing requirements for procurement, and this approach will ultimately lead to a greater degree of competition with established production chains and value chains.

A similar effect is derived from the initiatives of innovative competitors, these competitors' efforts to achieve global recognition, through the use of digital platforms for research, development, marketing, sales and distribution, can quickly overtake longstanding participants in the market and suppliers of goods, by improving the quality, speed or cost of the goods. Additionally, significant changes on the demand side are occurring, as the increasing availability of information, constant participation of the consumer and new patterns of consumer behavior (primarily due to the constant association with mobile networks and information) have made companies alter the way they develop, market and provide products or services. The creation of information platforms is a significant trend. These platforms combine supply and demand, and undermine existing production structures. As examples, we can observe new models of business in the contemporary economy that revolves around sharing and on-demand.

These platforms are simple to utilize on smartphones, they bring people, assets and data together, which creates entirely new methods of consuming goods and services. Additionally, they facilitate the achievement of "wealth" for businesses and individuals more easily, altering the personal and professional fields of employment. These new business models are quickly growing as a large number of new services are released, including: launder, shop, housekeep, and transport. Overall, there are four primary effects of the fourth industrial revolution on business that have an impact on consumers' expectations, products, collaboration, and organization. Tangible products and services can be enhanced through digital means that augment their value. New technology makes assets more environmentally friendly and versatile, while data and analytics alter the way they are maintained. The consumers' individual experience, which is easily accessible information and the efficient utilization of resources necessitates new methods of cooperation, specifically considering the rapidity of the changes. Additionally, the emergence of global platforms and other new business models has led to a change in the culture of society and the organization of business. As a result, the individual, the culture of society and the organization of business should be reconsidered. As biological, material, and digital worlds continue to interrelate, new technology and platforms will provide increasing opportunities for citizens to communicate with government agencies, to share their thoughts, to coordinate their efforts, and to avoid government supervision.

Concurrently, governmental agencies will have access to new methods of technology to enhance their control over society, based on a more intricate and advanced system of monitoring digital infrastructure. However, the authorities will have to reconsider their methods of interaction with the civil population and pursue policies, as the central role they play in the latter will be diminished by the emergence of new sources of competition, distribution and decentralization of power stemming from new technology. Ultimately, the stability of government structures and agencies will depend on their adaptability. If they can recognize the world of significant changes in architecture, if they can alter their structures and become more transparent and efficient, then they will address new problems. Otherwise, they may have troubles with difficult issues.

This will be especially apparent in the field of management. When the second industrial revolution occurred at the same time as the current public policy and governing systems, decision makers had the opportunity to consider specific issues and propose necessary solutions or appropriate regulations. The entire procedure was linear, mechanical, and had a strict top-down approach. This approach is not appropriate for our era. Because of the rapid rate and magnitude of the impact of the fourth industrial revolution, legislators and regulators have encountered unprecedented difficulties, and have primarily found themselves with limited assistance.

For instance, how could they safeguard the interests of consumers and the society as a whole while still promoting technological innovation and advancement? With the advent of "flexible" public administration, the private sector has adopted appropriate strategies to create software and business models that are more extensive in scope. It implies that those in charge of regulation should take into account a new, ever-evolving environment, they should become fully familiar with what they are regulating. To accomplish this, the authorities and regulatory agencies should collaborate closely with business and society civil.

Impact on Humanity

The fourth industrial revolution will have an effect on both what we do and who we are. It will have an effect on our identity and all other associated aspects: the way we perceive privacy, the understanding of property, our consumption habits, our time devoted to work and entertainment, career development, a set of abilities and knowledge, and personal relationships. The answer to this question is the constant use of smartphones that would lead to the loss of one of the most important aspects of our lives: a pause, a thought, and the start of a meaningful conversation. Privacy is one of the most significant personal difficulties associated with new information technology. We consider the importance of everyone to be paramount, despite knowing that tracking and sharing information about ourselves is integral to the new information exchange. The fundamental questions regarding the loss of control over our personal information in the near future will be discussed primarily. Alongside advances in biotechnology and AI that redefine the definition of the human being as a whole and push back the boundaries of life expectancy, cognition and human capabilities will also have an effect on our moral and ethical beliefs.

Artificial Intelligence as one of the main drivers of the Fourth Industrial Revolution

AI already exists in our lives, this is exemplified by the use of cars and drones that are autonomous as well as virtual assistants that can translate or invest. Because of the exponential increase in computing power and the vast amount of data that has been accumulated, the field of AI has experienced a significant increase in recent years. AI is commonly employed in the creation of new drugs and the development of algorithms that can predict what we'll like to do in the future. Additionally, digital production methods interact with the biological world. Engineers, designers, and architects utilize computer-aided design, additive manufacturing, materials science, and synthetic biology to explore the potential for symbiosis between microorganisms, our bodies, the substances we consume, and the houses we live in.

Before AI was commonly utilized in science fiction, but now it is one of the most exciting and rapidly growing technologies. Limited or "weak" AI technologies are already commonly employed in various areas: from cell phones and everyday electronics to military goods. The creation of "strong" artificial intelligence that can lead to informed decision making is currently in the works. Experts predict that the potential for creating this technology will question not only the current system of global labor distribution, but also the global order and security system.

Despite the rapid evolution of AI, the "black box" issue, when computers that make decisions remain unexposed to developers, continues to be problematic. As a result, before employing artificial intelligence solutions pertaining to the utilization of lethal weapons, their transparency must be increased significantly. However, there is a potential conflict of interest when it comes to combining the transparency of machine learning mechanisms with their safety from the adversary, as the data used by neural networks can be intentionally poisoned (data poisoning). Also, it's important to recognize that military organizations have a smaller base of data that can be utilized for machine learning than other organizations that are not military.

However, it's already obvious that in the medium term, AI will facilitate more efficient production, transportation and trade, increase crop yields, provide new opportunities for technology development, restructure the labor market and modernize the military system in a positive way. One tendency is particularly troubling: ultimately, it is impossible to ignore the possibility that countries will have to automate the decision regarding a retaliatory strike, this can lead to erratic results. Additionally, the situation is complicated by the fact that autonomous weapons and artificial intelligence are still considered a "gray" area of international law. It implies that countries that can create and utilize new technologies in the field of AI have a good chance of economic growth and enhanced national security. Conversely, countries that rely too heavily on antiquated infrastructure and economic principles will have a hard time maintaining competitiveness.

Chapter 2

Digital Authoritarianism

While digital technologies have provided governments around the world with a means to communicate with their constituents, understand the popular mind, assess the potential political consequences, and alter government policies, these same technologies have also given these types of governments an unprecedented capacity to remain in power. In fact, while surveillance, propaganda and false information have always been part of the typical political strategy, several technologies make this suppression and control more widespread, efficient and subtle. Nowhere is this more apparent than in China.

Digital authoritarianism manifests in multiple forms; online attacks, including harassing, disinformation, internet shutdowns, cyberattacks and targeted surveillance using social media, artificial intelligence and facial recognition technology. The methods employed by Saudi Arabia or a technological powerhouse like China are undoubtedly more advanced than those employed by the Filipinos, Egyptians or Iranians. However, lower-tech doesn't mean that there is a diminished threat. Governments that lack political, technological and economic resources are still capable of undermining democracy and human rights.

Additionally, it's crucial to recognize that elements of digital authoritarianism can and have been present in liberal democracies like Canada and the US. The truth is that the global decrease in democracy over the past 15 years occurred alongside the rise of new methods of information gathering, communication and surveillance. It's worth investigating this intricate connection. Autocratic leaders and nations have developed a tendency to utilize technology. They've learned that the survival of the regime can be dependent on their ability to utilize and deploy technologies against their own population. The concept of authority has been re-created: it is now digital.

While the digital revolution began in the 1950's, the emergence of new digital and information technologies, nearly 20 years ago, marked a new era for the technological revolution and had a significant impact on the digitization of society. Politics, society and markets have become intrinsically linked and dependent on one another in unprecedented ways. The potential for these technologies to affect economies and geography is disturbing. Early on, a general consensus was that digital technology, the internet, social media, artificial intelligence and machine learning would facilitate speech freedom, reduce power-ups and spread democracy across the globe.

Today, the majority of this technological utopia appears to have been lost. The country that has mastered the art of digital authoritarianism is undoubtedly China. For decades, Beijing has dedicated funds to a large technological system intended to regulate the Chinese population in all areas of their lives. Starting with the "Great Firewall" internet access was reserved for Chinese citizens and was heavily influenced by the government. China's perspective on cyber sovereignty is opposed to universal human rights. Also in Beijing, there are closed-circuit television cameras, sensors, and AI that are deployed across the country, this makes surveillance a common occurrence.

As mentioned in this chapter digital authoritarianism is not limited to illiberal governments. Indeed, democracies like Canada, the UK, India, and the US are not immune to methods in the toolkit of digital dictatorships, especially since the use of surveillance and tracking has become part of many technologies in our everyday environment. We must look within ourselves to protect the privacy and data of citizens at home while condemning techno-dictators, including those who perpetrate cross-border repression of activists living within our borders.

The growing threat of digital authoritarianism is well known in the West, the problem is well documented. However, this trend is not only continuing, but also accelerating in almost every country on the planet. Digital authoritarianism faces overlapping and growing challenges within authoritarian states as a tool to weaken adversaries through export to like-minded regimes and democracies themselves.

The exploitation of digital tools by authoritarian-leaning leaders threatens to leave humanity behind and threatens to undo the political and multilateral decades after World War II in every country, from the world's most important democracies to the most fragile progress. A range of information technologies, envisioned by their creators as a bridge to improved governance and freedom, actually suggest that their double sharpness runs deeper. From disinformation to industrial espionage to citizen surveillance and election meddling, the misuse of these technologies and the number of actors wreaking havoc in cyberspace is growing. At the same time, the internet is being deliberately fragmented, potentially to the advantage of authoritarian states. The flow of information is increasingly dictated at the national level, limiting the competition of ideas and suppressing the rights of individuals who must publicly follow the rules, while favoring those who have broken the law with few consequences.

Established democracies, while often preserving and promoting the beneficial elements of technology, lack a coherent and collective strategic approach to combating dictators' use of digital and online spaces. As a result, no concrete action has been taken to halt or reverse the destructive trends of digital authoritarianism.

Why digital authoritarianism is a threat to our freedoms

Digital authoritarianism is the use of the Internet and related digital technologies by leaders with authoritarian tendencies to reduce trust in public institutions, increase social and political control, and/or undermine civil liberties. Human rights and civil liberties are at stake, including freedom of movement, freedom of expression and political dissent, and the right to privacy online and offline. Digital authoritarianism usurps and undermines the fundamental principles of democracy and open societies; her goal is not simply to break them down, but to redefine and reshape their authoritarian image.

The battle between defenders and attackers in the digital space is inherently about trust. Public trust is arguably the most valuable asset of a democracy, the basis of the social contract between the state and its citizens. Without public trust, institutions collapse and even the most established democracies experience decline.

Digital authoritarianism presents four overlapping challenges.

1. It is spreading within existing authoritarian states such as China, Russia, Iran, and Saudi Arabia. Artificial intelligence, facial recognition, and the Internet of Things enhance the ability of authoritarian regimes to monitor and control individual citizens.
2. Authoritarian regimes are extending the reach of their digital tools overseas, increasing overt and covert surveillance of their own citizens and those of other countries. They also actively use these tools to weaken recognized state and non-state adversaries.
3. Digital dictators export their tools to like-minded regimes, both to strengthen ties with them and to profit commercially from them. Tools of control are also commercially exported from democracies to partially free or dictatorial states.
4. Political parties, interest groups, and private companies in democracies are adopting the tools, techniques, and tactics of digital authoritarianism at the expense of public trust, privacy, and other civil liberties.

Digital authoritarianism is a living experience in both democratic and non-democratic contexts, although it manifests itself differently in each context. As states with histories of authoritarian rule evolve and increasingly export their tools and strategies, democracies and their institutions are ill-prepared and vulnerable to internal and external corruption.

The four overlapping ways in which digital authoritarianism challenges the aforementioned principles of democracy and human rights are all disturbing and deserve better strategic responses. The fourth threat, however, poses the greatest, longest-term, and most immediate threat for the simple reason that while only democracies can stop digital authoritarianism, they are actively being consumed by it. Political parties, interest groups and even corporations in democracies have continued to adopt the practices of authoritarian regimes, especially over the past five years. Despite waves of democratic liberalization in the 20th century, the rise of digital authoritarianism heralds the opposite trend of this century, permeating dictatorships and democracies alike.

Politicians in major democracies adopted guidelines in domestic politics rooted in the authoritarian playbook of Leninism and Nazism. These actors seek to cast doubt on established institutions and non-aligned elites in order to gain power and evade scrutiny and accountability. They refuse to judge the legitimacy of any individual or institution by their record or the viability of their policies. Using digital tools, they are expanding their reach, creating powerful new information platforms outside traditional media that, once opened, can make bold and jaw-dropping claims better suited to attracting national attention.

With a strong communications foundation, they can quickly grasp or redefine the political landscape. They are also opportunistic, unafraid to seek support wherever they can, including extremists at home and sympathetic or malicious supporters abroad, who help in a variety of ways, from financial support to providing hacked information, to running disinformation campaigns in their countries represent. In some cases, these authoritarian-minded parties and politicians have proven better than others at exploiting commercial data collected by companies and suppliers on individual citizens. Their ability to combine disinformation with targeted advertising is a powerful political weapon in an already polarized environment.

A major obstacle to combating digital authoritarianism is the lack of public-private collaboration. It's no exaggeration to say that the tech sector has been a key driver of overall economic growth, at least in the United States. Yet the tech industry's growth in size and influence has coincided with a steady decline in internet freedom and the rise of digital authoritarianism. Reasonable questions have been raised about the responsibility of these companies in combating malicious activity and whether they have grown too powerful without adequate oversight. Internet communications in a free society are often controlled by private technology companies. Governments, especially elected officials, often lack expertise in rapidly evolving technologies. Digital authoritarianism undoubtedly thrives in environments with weak governance and weak cooperation between the public and private sectors.

There may be some developments in the future. First, COVID-19 has shown that it is accelerating existing trends, including the further concentration of digital power in the hands of governments at the expense of citizens' privacy. Mass protests are another escalating global trend, and even before COVID-19 and other civil society and grassroots mobilization efforts increasingly moved online, dictators had more digital tools to suppress them. What the COVID-19 restrictions have also done is provide unprecedented opportunities and excuses for autocrats to gain greater control over society both online and offline, something they are unlikely to give up easily.

In essence, digital authoritarianism, also known as technology-enhanced authoritarianism, is the use of technology by authoritarian governments not only to control the behavior of their citizens, but also to influence them through surveillance, repression, manipulation, censorship, and appropriate delivery of information services, all maintained and expanded within political control. Many illiberal or authoritarian governments in East and Central Asia, the Middle East, Africa, and Latin America use technology to control their citizens and support their regimes. But China and Russia are major practitioners, and China in particular is spreading its digital authoritarian models and mechanisms through a combination of technology exports, domestic example-setting, and international engagement.

Identifying, monitoring and censoring people online is the most blatant form of digital authoritarianism. But that's just the tip of the iceberg. Digital authoritarianism means much more than online censorship. This includes personal and mass surveillance in support of authoritarian governance through the use of cameras, facial recognition, drones, GPS tracking, and other digital technologies. It normalizes constant surveillance and removes expectations of privacy. Digital authoritarianism also includes governments' disinformation campaigns aimed at manipulating citizens, while punishing and censoring dissent on internet platforms and elsewhere under "fake news" laws.

In addition to centralized control over Internet governance and infrastructure, digital dictators support and co-opt their own tech industries to support their efforts to maintain social control and build state capacity to enhance their legitimacy. Digital authoritarianism includes technology-enhanced systems of incentives and punishments, such as China's social credit system, which institutionalizes the transfer of data between private technology companies and government agencies, allowing only compliant citizens to participate fully in society and the economy. While digital authoritarianism allows governments to exercise effective control and coercion over citizens, it also reduces the need to resort to coercion in the first place.

Digital dictators often employ the principle of cyber sovereignty controlling the internet within national borders. This goes against the fundamental principles of the Internet, namely net neutrality and the free flow of information. Cyber sovereignty enables governments to better control their citizens' information environments so that they no longer need to be censored. While digital authoritarianism allows governments to exercise effective control and coercion over citizens, it also reduces the need to resort to coercion in the first place.

The technology allows for enhanced self-censorship, social surveillance, and reporting on the behavior of fellow human beings through digital applications with less obvious, more automated controls, and more subtle ways. Control over the technological sector also allows for the collection and use of vast amounts of data that are fed into algorithms developed and used to shape economic and social interactions that promote the values and norms of the ruling regime

Digital authoritarian technology ecosystems generate vast amounts of data that states can access and analyze to shape and control society through surveillance, propaganda, and social credit systems.

Another function of big data collected by authoritarian regimes is to integrate with algorithms to develop artificial intelligence, which can not only monitor the whereabouts and online behavior of individuals, but also map their relationships through link analysis to determine their intentions or emotions, using detection Sentiment analysis and extrapolation of their past or future positions and actions to maintain regimes.

China is clearly the leader of a technology-driven authoritarian state whose model of digital authoritarianism includes the practice of many of the elements described in this chapter. The Chinese government is working with technology companies to intensify the enquiry, collection and processing of personal data at scale through online activity, from social media behavior to purchasing habits. With no choice, more than a billion Chinese use only a handful of mobile phone apps. Despite the convenience of these phone apps, users' communications, transactions and behavior are shared with big tech companies such as Ant Group and Tencent, which are obliged to share this data with the Chinese government.

Chinese people have become accustomed to bringing their personal hobbies, education and health, education, economic status, eating and consumption habits, social interaction, and even reading hobbies into the vision of large technology companies and the country. Combined with a large and rapidly growing fleet of cameras equipped with facial recognition systems and crowdsourced reporting systems, it enables unprecedented granularity of surveillance and allows manipulation of individual behavior. In Xinjiang, in northwestern China, for example, in addition to ubiquitous cameras, most residents are required to download apps on their phones so authorities can monitor what they see and track their whereabouts. Chinese authorities have been closely tracking the locations of more than 2 million people using facial recognition companies and a police service provider called SenseNets.

The Skynet Project is a video surveillance project invested and launched by the Chinese government in 2003. The government installs video surveillance equipment in public gathering places such as transportation hubs and security checkpoints, and uses GIS maps, image acquisition, broadcasting and other technologies to monitor and record real-time information in various areas. The Skynet system connects surveillance cameras in different places such as train stations, restaurants, shopping malls, theaters and other public places, buses, subways, taxis and other means of transportation, and can identify a large number of people in a short period of time. Chinese companies Hikvision, SenseTime, Huawei and ZTE have all participated in the construction of the Skynet project. The Skynet system has about 200 million public surveillance detectors in mainland China.

In 2015, the National Development and Reform Commission and the Central Political and Legal Committee launched the "Sharp Eye" project. The main goal of this project is to achieve comprehensive real-time monitoring of rural areas by building high-resolution cameras at main road entrances and crowded places in rural areas. Sharp Eyes also puts surveillance capabilities in the hands of citizens and encourages their direct participation. This project uses the existing rural TV network to connect public security video monitoring information to rural household digital TV terminals. From the perspective of the police, the purpose is to achieve "full coverage, full network release, ready availability, and full controllability."

As a result, Chinese authorities are integrating ancient and cutting-edge technology such as mobile phone scanners and facial recognition cameras into various tools of authoritarian control. For example, the face recognition system Skynet Project includes multiple application systems such as a face detection and comparison system, a face retrieval system, and a video retrieval system. These systems mainly use face recognition algorithms, face tracking algorithms, face quality assessment algorithms and face recognition algorithms, as well as personnel monitoring and motion tracking. Chinese police can access and search facial recognition databases in facial recognition servers, which perform facial image modeling and analysis. Chinese police can also deploy and install cameras at key surveillance locations to capture the faces of people crossing Skynet.

The Chinese Communist Party has always sought to enhance the regime's legitimacy by shaping public discourse, mobilizing its support base, and suppressing all political and social protest movements. The party-state actively subverts and hijacks social media for its own purposes. Now, it is also using algorithms, automation and human curation to purposefully spread misleading information, further increasing the effectiveness of its propaganda machine.

There are many types of "sensitive information" on the Chinese Internet, including so-called "inside information" such as propaganda bans and inside stories about high-level political struggles, as well as social issues such as corruption, housing prices and health care reform, wages and pollution.

The CCP also uses high-tech censorship systems, as well as official media reports to increase internal and external ideological propaganda, especially the manipulation of nationalism. The methods used include censoring information, distorting facts, altering narratives, and deliberately leading people to forget history. Social media accounts are supported or set up directly by government agencies, and deceptive digital tools such as bots, botnets and trolls are also common.

The COVID-19 pandemic originated in China. One of the reasons this outbreak has spread so widely is directly related to the CCP's control over the internet. During the pandemic, the Chinese government has deployed a variety of surveillance technologies, including tracking apps, drone surveillance, and cameras. Indoors and outdoors, mask wearers can be identified through remote temperature sensing and enhanced facial recognition. A mobile app called "Health Code" has good news for public health and a threat to privacy. When people scan and board a bus or enter a restaurant, they may be stopped if they have a bad review or are having trouble using their smartphone. The "Health Code" is rapidly gaining popularity in China. The Chinese are very willing to enter their data into the health code program every morning. The pandemic has become a longstanding excuse for the CCP's digital authoritarianism.

China's experience in using digital tools for home inspections and surveillance has made it the vendor of choice for illiberal governments that want to deploy their own surveillance systems. Chinese officials organized new media and information management training and seminars for representatives from 36 countries. Chinese state-owned and private companies are building telecommunications infrastructure in many countries around the world, and surveillance companies such as Hikvision and Yuncong

Technology are selling facial recognition technology using artificial intelligence in countries such as Egypt and Qatar.

China's response to the pandemic, aided by the expansion of digital authoritarianism, is an important part of its efforts to strengthen its position on the world stage. However, the Chinese Communist Party is under increasing pressure from its own citizens as well as the international community. The CCP has responded to these criticisms with further censorship and propaganda in its domestic and external diplomatic efforts. In China, for example, the Cyberspace Administration of China has banned videos or posts online that criticize police enforcement of quarantine restrictions. Social media posts critical of the CCP's handling of the outbreak were routinely blocked or deleted, as were calls from critics for greater transparency and accountability.

This online suppression of information is increasingly spilling over into the real world. There is evidence that China has stepped up punishment and detention of those who are critical or negative about the government's response to COVID-19 or who call for greater online freedom. Media reports published by Freedom House China have tracked multiple incidents of ongoing detention of critics and online censorship since the outbreak began.

The pandemic has already shaped the future. Its many legacies include the normalization of technological surveillance, the concept of a surveillance state, and China's opportunistic advancement of its technological agenda. Democracies have a responsibility to do more to address the disruptive aspects of China's global technology agenda and to find ways to harness technologies, including artificial intelligence, for the benefit of the global public good, while preserving hard-won democratic rights and freedoms.

Chapter 3

The Rise of a Cashless Society

Money has always been unable to buy happiness, but today money is becoming more and more virtual and intangible. This means we use digital accounts to represent money, not coins and notes. Some countries, such as Belgium, have made it illegal to buy real estate with cash. In Australia, some banks have moved away from cash entirely, while in Sweden less than 1% of payments are made in coins or notes. In response to this trend a new law in the US was created, the Cashless Retailers Prohibition Act 2018. This law would make it illegal for businesses such as restaurants and retailers to refuse cash or charge different prices based on how customers pay. The aim is to ensure that everyone can participate in our cashless society, no matter how they prefer to pay. Despite a lot of talk about mobile payments and digital currencies, cash is still widely used, especially in countries that require it. In Latin America, some 130 million people work informally and rely on cash for goods. They find it difficult to get loans from banks because they have no credit history. Therefore, they can only rely on cash for shopping. If cash is not an option, they look for alternative payment options, such as using someone else's credit card or an installment plan. Not only do you need cash to buy clothes, eat out, etc., but you also need to pay utility bills such as water and electricity. Brazilians, Mexicans, Colombians and many other Latin Americans pay their bills in cash using local cash payment methods such as Bancário, OXXO vouchers and Via Baloto.

In Latin America, people use different cash payment methods that are deeply embedded in their culture and also allow them to shop online. Many people are afraid of online scams, so cash payments are a suitable option for them. When people are empowered to use payment methods they can access and prefer a more inclusive payment system will emerge, benefitting both consumers and businesses. For example, EBANX has seen its customers in Latin America triple their sales by accepting local payment methods. By refusing to accept cash payments, companies not only discriminate, but deny sales and customers who have the will and money to buy from them. While some parts of the world are rapidly adopting cashless payment methods, others may need to take a step back to give the rest of society time to catch up.

Governments are moving towards a cashless society

Governments and banks around the world are pushing for a cashless society. However, this evolution to digital transactions means that individual liberties are being compromised as all transactions are monitored and controlled by those running the technology. In a cashless society, every purchase has an electronic paper record, and debt can be paid off with the touch of a button.

While it's convenient to pay bills or shop online without having to carry cash, paper money could suddenly be obsolete, rendering everything people save worthless. Furthermore, banks and governments can decide at any time that individuals owe them money and keep debiting their accounts without giving them a chance to fight back. With cash, people have the ability to hold or save their money, but this is not possible in a cashless society.

This book will discusses major problems for many who rely on cash for transactions and highlights the possible downsides of a sudden transition to a cashless society.

Governments are shifting towards cashless systems

The Indian government took a step towards a cashless society by announcing that 86% of the cash in circulation cannot be used for transactions. The move, known as "demonetization," aims to curb tax evasion. As a result, people are scrambling to deposit Rs 500 and Rs 1,000 notes in banks, creating difficulties for the poor who rely heavily on cash transactions.

The bigger problem with this move, however, is the loss of personal freedom and financial sovereignty. In a country where most transactions are settled in cash and many people are unbanked, the government is now forcing people to opt for banking services, giving more power and money to a corrupt banking system. Governments and banks want a cashless society where every transaction can be monitored and all debts and taxes can be easily settled. This means that if the government or the bank claims that someone owes money, they can immediately withdraw the money from their account without waiting for payment or hassle. While officials claim they won't abuse their power, history shows corruption in the banking industry, such as illegal foreclosures and the creation of fake accounts to meet sales quotas, making the prospect of a cashless society worrisome.

Banks are adopting cashless methods

Big banking groups have already begun microchip implants on credit cards, and it is possible to take direct stakes in human microchip facilities. Barclays is teaming up with Japanese tech firm Hitachi to develop a biometric reader that scans a finger and recognizes a unique pattern of veins to access accounts, rather than using passwords or PINs.

Human microchips are being hailed as the next advance in biotechnology, with the next decade set to be marked by "virtual reality, implants and transhumanism". Overall, moving away from cash has potentially negative consequences for people who rely heavily on cash, especially in developing countries. Taking Latin America as an example, about 130 million people are engaged in informal work, mainly relying on cash to buy goods, and because they have no credit history, it is difficult to obtain credit from banks. Additionally, some countries have introduced cashless payment methods and others are considering them, such as Australia and India.

The shift to digital transactions means that personal freedom is being compromised, as all transactions are monitored and controlled by technology operators, and debts can be repaid at the touch of a button. The transition to a cashless society also means that people will no longer have the ability to actually hold or save their money, leaving them vulnerable to exploitation by governments or banks.

How soon will the world be cashless?

The decline in importance of traditional forms of money is having a major impact on trading practices in countries around the world. Let's take a look at the benefits and potential downsides of a cashless economy. A world without money and paper money as a means of exchange seemed unimaginable ten years ago.

The most technologically advanced among us will no doubt see the day when traditional forms of money become obsolete. However, since the vast majority of the world takes money very seriously, this seems to be nothing more than a pipe dream. Although most purchases were made after the shift to debit and credit cards in 2012, cash still dominates. This is the case even though most of the deals were completed in 2012. Then COVID-19 came and suddenly everything was different.

What Would a Cashless Society Look Like?

Humans have been interested in evolution since the beginning of recorded history. There is always a feeling that everything can be done in a better way, whether in terms of speed, efficiency, cost or safety. Pigeons are no longer necessary for us to deliver messages, as modern technology makes this possible. Now one company can provide thousands of customers with the items they need every day.

However, who would have thought that advances in digital technology would one day push cash payments to the brink of extinction? There is no doubt that younger generations use cash less than older generations. As mobile phones become an integral part of people's lives, people are used to meeting deadlines. Therefore, digital payment is much more convenient. Depending on how you look at it, we are getting closer to a "cashless society" which can be both exciting and frightening.

Reasons why we need to keep cash

Without cash, consumers are at the mercy of commercial banks and payment providers. With cash as an option, people can easily withdraw their money and store it elsewhere. Without cash, that choice disappears, as does the balance of power it creates between the citizen and commercial banking sectors. Cash is a vital pillar preventing governments and companies from monitoring every transaction and recording the movement of goods and the daily movements of people. In some cases, cashless payments are registered and used to profile and "score" citizens to control them. Cash is not used to profit from its use, but to support and maintain free transfers. There may be costs associated with cash, but cash itself is an instrument of value transfer that settles no more and no less than face value. Market economic conditions apply to all other payment methods; they must generate profits for their issuers. Cash is the only payment method that can be used unlimitedly, no matter when and where you pay. No bank account, good credit, cash payment requires internet access. Cash is the only form of payment that is independent of the issuer without the involvement of a third party.

Risks of a Cashless Society You Should Know

Due to the widespread spread of the coronavirus, many people are choosing to pay for goods and services using credit cards and other contactless payment methods instead of cash. This is done to minimize the possibility of close contact with others. Some companies are even mandating contactless payments and banning banknotes and coins for hygiene reasons.

1. The profits of small businesses are diminished

When paying by credit card or digital platform, the store is charged by a third party, which costs the company. These costs can add up quickly, which is a problem for many businesses that are just getting started and trying to thrive.

2. Digital or card payments put your privacy at risk

When you swipe your credit card, all the companies you do business with have access to information related to your payment.

3. This Increases Your Exposure to Data Breach

Leaking sensitive information increases the risk of a data breach. If your credit card information falls into the wrong hands, criminals may use your credit card information to make unauthorized purchases. In the event of identity theft, more sensitive data could be stolen if retrieved.

4. You become more dependent on technology for all transactions

Cash is the most trusted method of payment unlike other payment methods which are not always as reliable as cash. This could be because your credit card was declined or your phone battery died and you can no longer make payments through the app on your device. The success of cashless transactions depends on technology, which is sometimes unreliable.

5. Managing your finances could be more difficult

Some personal finance experts recommend paying with cash instead of credit cards or other payment methods, both to raise awareness of your spending habits and to ensure you're not spending more than you actually have. Swiping a credit card or using a cash app is often so indiscreet that users don't really think about what they're spending, which can lead users to spend more than they can really afford.

6. Transaction Fees May Be Applied to You

It's not just the business you buy from that needs to pay for cashless interactions; you may also need to pay to live in a cashless society. For example, peer-to-peer payment platforms like Venmo charge fees for certain types of transactions. Credit cards have their own fees, such as B. annual fee, late fees, and transaction fees for other countries.

7. It could result in banks imposing higher fees

Individuals are likely to rely on credit cards and apps for payments, which could lead to negative interest rates and lower revenues for banks. In response, banks may increase fees, which end up costing customers more money.

8. You can no longer select your payment method

Some items you'd rather pay for in cash. As we move towards a society where money is not used as a medium of exchange, you will no longer be able to pay with cash.

9. It takes away money's emotional value

You've probably owned or had a piggy bank at some point in your life, and you probably love dropping your pennies into the little slot of the piggy bank. One could argue that once money is no longer physically present, it loses its ability to evoke strong emotions.

Chapter 4

The Social Credit System

The Social Credit System is a public credit rating and blacklist that the Chinese government is developing. This action is intended to track and assess the liability of companies, individuals and government agencies using a system of record. Several versions of the system are being experimented with, but for the general public, the non-surveillance system is based on blacklists and whitelists. Contrary to popular misconception, the programmes is primarily enterprise-focused and decentralized rather than centralized and individual-focused.

The system's roots date back to the 1980s, when the Chinese government attempted to develop a special banking and tax credit system for herders and small businesses that needed proof of records. In the early 2000s, the project was inspired by foreign credit scoring systems, and began to be piloted in China in 2009. It was established in 2014 as a public company with eight credit rating companies and was officially launched by Chinese Premier Wen Jiabao in 2011 during a meeting of the State Council. The social credit system is an extension of China's tax credit system and is jointly administered by the National Development and Reform

Commission, the People's Bank of China and the Supreme People's Court. The system should regulate credit functions and perform fiscal and social evaluation of companies, government agencies, individuals and NGOs or non-governmental organization.

The Chinese government aims to use the system to boost social trust and regulate businesses in areas such as food safety, intellectual property theft and tax fraud. Nonetheless, there have been major misreportings and misunderstandings about the social credit system in the mainstream media, including widespread misconceptions that Chinese citizens are rewarded or punished based on numerical scores assigned by the system that their opinions are taken over by artificial intelligence, and that Chinese citizens are constantly under surveillance. These misunderstandings are largely due to repeated offenses, grandstanding, conflicting information, and a lack of thorough analysis. The system likely began in the 1980s, when the Chinese government attempted to create a personal banking and credit rating system, especially for rural populations and small businesses that lacked records. The scheme first emerged in the early 2000s and has inspired credit rating systems in other countries. The initiative launched a regional test in 2009, before launching a national test involving eight credit rating firms in 2014.

The social credit system is an evolution of the current credit rating system in China. Overseen by the National Development and Reform Commission, the People's Bank of China and the Supreme People's Court the system is designed to conduct financial and social assessments of companies, government agencies, individuals and NGOs. The Chinese government hopes to boost public confidence in the country through institutional and corporate regulations. These challenges include food safety, intellectual property theft and financial fraud. However, the social credit system has led to a lot of inaccurate reporting and misinterpretation by the media due to language errors, grandstanding, inconsistent information, and incomplete research. The widely held belief that the Chinese reward and punish based on points provided by the system, artificial intelligence making judgments, and constant surveillance of Chinese citizens are just a few examples of these fallacies.

History

In the early 1990s, China developed a credit rating system for individuals and banks. It was then that the social credit system was first born. Western business credit systems such as Fico, Equifax and TransUnion have influenced it. The credit system is designed to facilitate financial assessment in rural areas, where people and small businesses often lack financial records. In 2007, government agencies formulated the first social credit system plan. The social credit system also seeks to close the moral loopholes, inadequate market regulation and income inequality created by the acceleration of economic and social development since China's economic reforms in 1978. These challenges have raised trust issues in Chinese society, including scandals related to food safety, violations of labour laws, intellectual property theft and corruption.

The concept of the social credit system can be traced back to the legalist meritocracy and promotion system adopted by Shang Yang during the Warring States Period to improve the efficiency of the Chinese state. The politics of the social credit system stem from policing and labour management practices. The modern Chinese government still maintains a personal paper record system. However, due to the greater difficulty of putting paper records together for quick and thorough analysis, these systems cannot provide Chinese citizens with the same level and speed of feedback and results as comprehensive electronic systems.

The grid social credit system is a police method that was first adopted in a few places from 2001 to 2002 under the leadership of Jiang Zemin, General Secretary of the Communist Party of China and is the origin of the social credit system. Jiang Zemin's government proposed in 2002 to establish a social credit system to promote a "unified, open, competitive, and orderly contemporary market system." In its early stages, grid policing served as a framework for improving communication between public safety agencies.

The grid approach was quickly adapted for social service distribution. Grid management improves the tracking and surveillance of individuals and situational awareness of group-level authorities. However, sociologist Zhang Lifan pointed out in 2018 that Chinese society still lacks trust. Even if they are innocent, people think they will be used or get into trouble. In his view, it was caused by the Cultural Revolution, in which millions of Chinese were massacred and deliberately instigated among relatives and friends. The Social Credit System aims to restore trust among Chinese citizens.

Initial testing

State media now calls the list the first step toward a national social credit system, the Supreme People's Court in China is also working with commercial companies. Sesame Credit, for example, began withdrawing credit from customers who failed to pay court fines on time. In 2015, the People's Bank of China approved eight enterprises to carry out pilot social credit systems. The government has partnered with some companies to develop loan calculation software and algorithms.

Chinese private companies that have received government approval to conduct social credit experiments have established commercial pilots. The pilots are volunteers, have a wider scope than local government departments, and are always open to inquiries from the public. Customers with high scores will receive benefits including faster access to credit, lower car and bike share rates, faster visa processing, free health checks and special hospital care. The method used to determine scores in commercial pilot programmes remains a mystery, although some pilots use big data analytics and artificial intelligence methods. The social credit system was originally intended to be run by a private company, but in 2017, China's central government recognized the need for third-party management.

Unfortunately, no licenses are issued to private companies. The Chinese government has decided not to approve the pilot project as an official credit reporting system until mid-2017. Conflicts of interest, ongoing government oversight, and a lack of collaboration in data sharing among the companies involved in the development are some of the reasons. However, it is not excluded that social credit systems operate under the control of apparently external organisation, such as legal partnerships of trading companies.

How does the social credit system in china operate?

How does the National Trust system record the good and bad deeds of citizens? Data for social experiments in China are primarily drawn from online sources such as social media interactions, search histories, and shopping patterns. Everyone can influence their reputation and be monitored based on their online behavior. Other data sources include government agencies, nonprofit organization data, criminal records, and financial organisation. Finally, China intends to monitor a person's true behavior through the video surveillance currently deployed. Advanced facial recognition technology will monitor people's movements and interactions with their surroundings.

Which acts impact citizen's credit score?

Each citizen's score starts at 1,000, but various actions can increase or decrease it from that point on. The minimum and maximum scores are 600 and 1,300, respectively. What behaviors bring you closer to these extremes? Much behavior depends on how one fulfills one's obligations to society, as the Chinese government and traditions have traditionally placed great emphasis on civic responsibility.

Good works

Donating blood, volunteering, caring for the elderly, and safe driving habits are all good activities to improve the status of civil society. Additionally, investing in literature instead of video games and setting aside time for education are two other ways to improve grades.

Bad acts

Negative behavior that damages a person's reputation includes defaulting on obligations, reckless driving, sharing information critical of the government on social media, spreading rumours, adultery and excessive purchases of video games.

What are the repercussions?

The higher a person's social score, the more benefits a person can receive. The more flaws there are the less valuable they are. The Chinese government often uses the following methods to reward and punish its citizens based on test results.

How are virtuous acts rewarded?

Good deeds are not immediately rewarded. Conversely, residents with high scores, indicating a disproportionately higher proportion of good activities than bad, can reap a range of benefits. Benefits include lower electricity bills, access to college and higher education, easier access to public housing, free access to gyms, and a higher likelihood of getting loans.

In addition, certain high-scoring residents enjoy tax benefits, reduced public transportation costs, and can book hotels or tours without a deposit. Putting these benefits into good activities can improve the overall happiness of residents, because China is one of the most miserable countries in the world.

How are negligent acts punished?

Citizens are more likely to be penalized for misbehaving when the score is below the starting point of 1,000. Residents with low housing standards may not have equal access to social services or amenities, and are less likely to have access to credit. Sometimes penalties are applied to crimes that lower an individual's score on the social credit system. For example, some people are confiscated when their dog doesn't clean its canines, or walk off a leash while walking the dog, which will deduct points. Some of the worst incidents resulted in public shaming on television, social media, billboards and other sites.

A Comparative Analysis of the Western and Chinese Social Credit Systems

China's social credit system is a programme that monitors and evaluates citizens' behavior to determine their social credit score. This value affects a person's ability to access certain privileges such as bank loans, faster internet, and air tickets. The system was originally developed by the Bank of China in the 1990s to increase lending in rural areas. However, the Chinese government quickly realized that it could use similar procedures to gather information about the behavior of its citizens. In 2014, the State Council issued a draft for soliciting comments on the first phase of the social credit system.

The government has run pilot programmes in 43 cities across the country, which Rongcheng uses to flag certain behaviors as acceptable or unacceptable. Every adult is assigned 1,000 social credit points and earns or loses points based on whether their public and private lives meet state standards. The system is designed to encourage good behavior and tackle everything from crime to bad driving to financial crime. However, the system has outgrown its original purpose. Chinese authorities have drawn up extensive blacklists of those deemed "unacceptable". There are 23 million Chinese nationals on those lists and counting. The system has been used to punish those who speak out against government corruption or censorship. Journalist Liu Hu, for example, was banned from flying, traveling on trains, buying real estate, and taking out loans without due process.

Muslim Uyghurs in Xinjiang are particularly affected by the social credit system. Many were sent to re-education camps for "unacceptable" behaviour. This has raised concerns about the future of the system and its impact on citizens' lives. Some worry that loyalty to the Chinese government will determine every aspect of its citizens' lives, from where they live to their children's education. Similar programs are being developed in other parts of the world. For example, countries such as Australia, Austria, etc. have introduced or attempted to introduce mandatory vaccinations. The social credit system and other similar programs have raised concerns about government encroachment and loss of liberty.

The US has also introduced programs similar to social credit in various contexts. For example, some US rating agencies use similar models to assign credit scores based on factors such as payment history and financial habits. Additionally, some employers use social media to screen job applicants for inappropriate or controversial online behavior. In the "Black Mirror" episode, Lacey's obsession with boosting her rating to avoid negative consequences ultimately leads to her downfall. This situation could become a reality for citizens around the world if we are not vigilant. It is important to understand these programs and their potential impact on our lives and our freedoms. However, China's social credit system remains the largest and most controversial implementation of the technology. As the system is phased in, many Chinese citizens have expressed concerns about its impact on their daily lives, including potential loss of privacy and increased government scrutiny of their behavior.

Implementing Social Credit Systems during the Pandemic

The current human age presents unprecedented opportunities for technological totalitarianism to enslave people around the world. Health passport regulations have been a turning point in human history as they raise public health and social control concerns. While these health passports make little sense from a public health perspective, they do appear to make sense for promoting authoritarian social credit systems through digital identity systems. COVID vaccines are designed to minimize symptoms but have not been proven to prevent transmission. Mandatory use of COVID passports therefore appears to have less to do with public health and more to do with social control.

Some leaders have publicly stated that vaccination is a prerequisite for full participation in society. Countries around the world have implemented draconian measures including vaccination passports, lockdowns and military intervention, leading them to become police states. By coercing, manipulating, and incentivizing people to show their papers everywhere, governments around the world began to emulate Communist China.

The US is moving in a similar direction. All over the world, citizens are rising up against tyranny and fighting to protect the most basic of human rights freedom. Protests have erupted across Europe, with millions taking to the streets against lockdowns, mandatory vaccinations and vaccination certificates. The flawed concept of immunization certificates as a public health measure is Orwellian in the way they are issued, and they are setting society down a dystopian path. According to the World Health Organization, major uncertainties remain about the effectiveness of vaccinations in reducing transmission, as well as the limited availability of vaccines. The situation we face globally therefore represents a turning point in history, and we must be very vigilant about the impact of these technological tools.

Privacy and freedom considerations in establishing a national ID card system

A social credit system is a way of tracking and evaluating a person's social interactions, demeanor. The system gives points or ratings based on a person's behavior and has the power to reward or punish. China is adopting the idea of a social credit system as a weapon of social control. Many other countries have also introduced social credit systems. On the one hand, government officials will argue that the social credit system brings many benefits. For example, rewarding people for good deeds can encourage them to behave in a more socially responsible manner. This could lead to lower crime rates, more civic engagement, and an overall more peaceful community. Social credit systems can also be used to identify and address problems in public services, such as inadequate health care or public education. However, there are serious problems with using the social credit system. The potential impact on liberty and privacy is one of the biggest. Such systems can collect vast amounts of personal data, including not only social media posts and online activities, but also behavior and activities.

This can significantly affect people's freedom of speech and opinion by giving the impression that everything they do is being watched. Another concern is that social credit systems could be used for political purposes and ratings could be manipulated in favor of specific individuals or organisation. In addition, subjective and potentially biased behavioral assessment criteria can lead to unfair treatment of individuals. Overall, social credit systems have the power to encourage good behavior and improve social outcomes, but there are deep concerns about the potential impact on privacy and liberty.

In China, people with low social credit are not allowed to buy plane tickets, which helps prevent people from leaving the country without paying off their debts. The Chinese Communist Party would argue that this type of intervention could help protect consumers from unscrupulous lenders and other financial predators.

The disadvantages of Social Credit Systems for Privacy and Personal Freedom

One of the main disadvantages of social credit systems is that they can be used as a tool of social control. By punishing those who engage in behavior that the government deems undesirable, social credit systems can be used to limit individual freedom and encourage compliance. This can lead to a lack of diversity in thinking and ultimately a lack of dynamism and creativity in society. Another disadvantage of social credit systems is that they may target specific groups of people. Additionally, social credit systems have often been criticized for their lack of transparency and accountability. The algorithms used to assign ratings are generally kept secret, and individuals have little recourse if they believe their ratings have been unfairly biased. This lack of transparency can lead to a lack of trust in the system, as individuals may feel that their fate is determined by processes they do not understand or have adequate control over.

Then there is the social credit systems negatively impact on privacy. By tracking people's behavior and social interactions, governments can collect vast amounts of personal data, including sensitive information about individuals. This allows for the creation of complete profiles that can be used for purposes other than assessing social credibility.

Impact on privacy and personal liberties

The establishment of the social credit system has raised concerns about privacy and personal freedom. Social credit systems are based on the collection of large amounts of personal data, including sensitive information about individuals. This data can be used to create a comprehensive profile of an individual that can be used for purposes other than social credit assessment.

This can lead to privacy breaches and a lack of control over personal information. Social credit systems can also have a negative impact on personal liberty. By linking social credit scores to social interactions and behaviors, governments can limit individual freedom and encourage compliance. This can lead to a lack of diversity in thinking and ultimately a lack of dynamism and creativity in society. To sum up, the social credit system has both advantages and disadvantages. While they can encourage good behavior and reduce administrative costs, they can also be used as tools of social control and negatively impact privacy and individual liberty. It is important that governments carefully consider the potential impact of social credit schemes and ensure that they are implemented in a transparent manner.

Aware individuals take social credit less readily

In summary, China's social credit system is not without controversy. Censorship prevents the Chinese from understanding the risks associated with the new social credit system. However, when they learned of the potential danger, their enthusiasm visibly waned. More research is needed on how the Chinese public perceives social credit, especially as new policies and frameworks continue to affect the lives of more and more people. Aspects of the social credit policy framework that require further investigation include public perceptions of metrics, such as people's perceptions of regulations that apply not only to "social credit offenders" but also to the offender's family and friends.

Inside China's Experimental Social Credit System

Rongcheng residents can check their social credit score at one of these agencies, to improve the "credibility" of China's economy and society, the government is testing a social credit system that combines Western-style credit ratings with broader and stricter standards. Online payment services rank alongside communities and individual companies.

The rich enjoy benefits such as cheaper heating bills and better access to bank credit, while the poor do not enjoy the benefits of modern transportation. Others fear it is just another step in further restricting China's already limited powers, as the Chinese Communist Party expands its grip on city governments and company boards. Unfortunately, since the advent of social credit, it has become harder to discern whether the promise and threat of the future is just that. Rongcheng is one of more than three dozen testing systems that have been deployed in cities across the country. The Chinese government and the scientific community have hailed it as an example of the system working as intended. It was revealed that their motives were not as clear as they had hoped.

Each of Rongcheng's adult residents initially gets 1,000 points. At this point, we can start crunching numbers. If you receive a ticket, five points will be deducted from your driver's license. Add 30 points to your score when you are recognized by your local government for doing something noble, running a great business, or being a big help to your family. If you take home a trophy belonging to your category, it's worth 5 points. The city's programme also recognizes time dedicated to volunteering and community service activities.

A resident's final grade can be anywhere from A+++ to D. Some violations may negatively affect the overall grade. For example, if someone is caught DUI, their grade will automatically be downgraded to a C. Even for businesses, social credit standards are strict. With countless scams and food scandals, Chinese residents have a reason to dislike companies and service providers.

To maintain your credibility in this market, it is important to pay your taxes on time and avoid fines for things like selling defective or dirty products. Companies with high scores are more likely to bid for government contracts and benefit from more favorable credit agreements. Although the system was introduced back in late 2013, many in the city are still unaware of its existence. People only realize this when they visit the gleaming town hall halls to apply for a government home loan or that coveted university degree, all to fulfill their wildest dreams.

Chapter 5

Digital ID and Biometrics

Governments around the world are designing or implementing digital identification systems, often involving digital IDs. The adoption of these systems is driven by a new development consensus that holds that digital IDs can promote inclusive and sustainable development and are a prerequisite for the realization of human rights. However, a specific digital ID model was promoted, drawing heavily on India's Aadhaar system as a source of inspiration. This digital identity system aims to provide individuals with a "transactional" or "economic" identity by establishing their uniqueness. The promise is that people with this economic status can do business with government and private actors. This will improve access to public and private services, boost economic growth, and contribute to the emergence of a truly digital economy. Unlike traditional civil identity registration systems, such as birth registration, this new model of economic identity sidesteps common conundrums about the legal status of registrants. Many see the rapid and massive deployment of such digital identity systems as dangerous.

Evidence of actual and potential, often serious and pervasive, human rights violations associated with this digital identity model has emerged in many countries around the world. Such systems can exacerbate forms of exclusion and discrimination that already exist in public and private services. The use of new technologies can lead to new forms of harm, including biometric exclusion, discrimination, and many of the harms associated with surveillance capitalism. The promised benefits of such a system have yet to be convincingly demonstrated. These dangerous digital identity schemes can cause "pain without gain." The World Bank, through its Identity for development initiative, is playing a central role in promoting and supporting this new digital identity system globally.

It played an important role in forging consensus in the newly defined field of "identity for development". By framing the problem as one among the "invisible billion" without official identities, and presenting digital identity systems as inclusive and empowering solutions, it empowers and empowers these systems. This is achieved through extensive publications, technical assistance to governments and knowledge sharing including many opportunities to learn from India's Aadhaar experience. Governments in countries such as Nigeria, Mexico, the Philippines, and West Africa are receiving substantial funding and support from the World Bank to develop and implement digital ID systems that represent this new economic model.

However, the World Bank is far from the only institution supporting the "identification for development" agenda. While the bank appears to be playing a vital role in rolling out these initiatives globally, a diverse network of stakeholders with diverse interests has emerged to support this digital ID model. These include governments, foundations, suppliers and consultancies. Despite ostensible references to human rights and development goals to legitimize these activities, there is currently an almost complete absence of serious engagement with human rights norms and values within this global network. Given the magnitude of the actual and potential human rights abuses caused by this digital identity model, we must not only sound the alarm, but also consider what can be done about it.

One thing is clear about digital identity systems: they raise serious human rights concerns and are prone to failure to implement. Even those pushing the "identification for development" agenda recognize these significant risks, however, evidence gathering and monitoring of human rights impacts are still lacking.

In India, the significant impact of Aadhaar on people living in poverty is only known through the efforts of academia, journalists and civil society organisations. This patchwork of evidence suggests that digital identity systems can lead to a variety of pressing human rights concerns, including but not limited to: civil rights violations; restricted access to healthcare, food and social security; various privacy and data protection, surveillance and cybersecurity concerns and fundamental changes in patterns of democracy, participation, and citizen-state relations. The human rights consequences can be severe and irreversible.

In India, for example, exclusion from the Aadhaar system has resulted in countless deaths from starvation and countless other situations of deprivation, exclusion and injury. Some of these negative impacts are not necessarily related to the digital aspects of such systems, but reflect underlying dynamics of social exclusion, economic inequality and marginalization. Any form of identification system can be used in both beneficial and harmful ways.

Digital ID systems can modify or amplify these impacts and counteract hard-won human rights gains. Other negative effects appear to stem directly from the introduction of new digital technologies and new methods of designing and implementing ID systems. This includes the use of digitized biometrics and platforms that centralize or centralize data for public and private use. For example, at the most basic level, the widespread use of biometrics creates new dependencies on information and communication technologies and often non-existent electrical infrastructure.

Many new or upgraded digital systems are also designed to encourage functional sneakiness, as they are intended for multiple purposes that were not foreseen when the system was first conceived. This means that damage can spread and intensify quickly as digital identity systems become an insurmountable barrier to widespread services and rights. At the same time, the ID4D initiative and other players in the field of identification for development have highlighted the potential benefits of digital IDs for development and human rights. Digital identity systems are designed to achieve a dizzying array of development goals.

This includes:

- Establish legal rights, sometimes including citizenship or refugee status.
- Facilitating access to public and private services.
- Improve financial inclusion.
- Build government capacity and statistics, open up new opportunities for the private sector and manage elections.

However, there is little concrete evidence to support these claims, backed by in-depth research and systematic monitoring. We don't know if these systems work. We still lack baseline studies showing unresolved problems, a reasonable assessment of whether new or improved digital identity systems can address any of the documented problems, and whether there are alternative, less harmful options to address them result.

The World Bank itself highlighted some of these evidence gaps. A causal link between digital ID systems and the resulting benefits is unproven, but still questionable and unproven. In many cases, these systems simply don't work. Many face serious surgical procedures Challenge and fall into corruption and waste. For example, an early attempt by the World Bank Introduction of biometric ID in Bangladesh greatly delayed after failed implementation Necessary legal reforms were made, but the project missed its goal of more than 50% population.

Combination of human rights abuses and business failures led to World Bank There simply isn't any solid "valid evidence" let alone enough to justify it the speed and enthusiasm with which digital identity systems are evolving. These evidence gaps on broader trends, World Bank figures are far from science Submit evidence. Other knowledgeable observers have not only emphasized the specific practice of banks is also the continuous lack of investment quality Impact assessments, especially of multi-sectoral work, and the lack of robust cost-benefit analyses.

Ongoing uncertainty also requires us to resist the urge to invest and massively accelerate an untested, unproven system in the entire population. However, proponents of digital identity remain aggressive Claims that digital ID systems represent a game-changer compared to paper-based alternatives Improved and should expand rapidly. In other words: compared to the counterfactual scenario we are better off investing in digital identity systems. However, this position ignores many very some of the world's most marginalized people are already suffering real harm. It also ignores the damage that such systems may cause in the future. Who is the promotion and financing of the rapid introduction of digital identity systems are also responsible for this Monitor and document human rights impacts human rights organisation and civil society groups will continue to play an important role in collecting and presenting evidence of human rights violations.

However, these organisations are already severely underfunded and face challenges; not and least of all is the closure of global civic space that threatens their very existence. This is It is unrealistic and negligent to place the burden of proof primarily on civil society, especially since relevant data on system design and implementation are not publicly available. How did we get to the point where digital identity systems violate human rights?

Promoted globally and attracted millions of dollars in funding without justification advantage?

It is helpful to first understand the broader field of "identity for development" its cast. Banks cannot drive these developments alone; many other actors, developers, including national governments, have invested heavily in biometrics Technology long before the development agenda was set last year ten years. Early approaches were driven by specific questions in different governance domains develop. For example, targeted, conditional money transfers preferred by many developing facilities requires complex approval criteria and vast amounts of data. These programs are said to suffer from widespread fraud, bugs and data breaches although reliable evidence of its existence or extent is lacking.

New technologies, especially biometrics, considered a "breakthrough" solution to these problems. However, the global flag because the development agenda is not just about applying new tools to old problems. This is It is also a well-crafted agenda for inclusion, empowerment and human rights. There are this is a "collective paradigm shift" that requires investment in new forms of systems replace by providing birth certificates, these new systems will help create a "digital public infrastructure". Part of the "digital stack" to "enable paperless, cashless, remote and data-enabled transactions". In other words, the new system will go far beyond the original capabilities of the legacy system this tightly integrated with the model of Aadhaar, the biometric unique identification project launched in 2011 India in 2009. Aadhaar provides a multi-purpose identification system, initially aimed at adults. This can be used to verify transactions with various public and private actors.

One of the main reasons why Aadhaar is admired by the developer community is the speed at which he works as a result, almost nationwide coverage has been achieved and currently reaches more than 1.2 billion people. Of course Even in India, the recognition gap persists and efforts are being made to reproduce Aadhaar is restricted in other countries. Identification systems rarely reach more than 80% of the total cover.

Basic ID systems do not need to be linked to legal recognition of rights; such systems may or may not be associated with legal status.

Instead, the underlying system promises in their neutral ability to distinguish between different people or to determine uniqueness, and use this as the basis for secure transactions. These systems look very attractive, because the problem after who is granted citizenship and the rights associated with it occur after identity verification system. This differs from birth registration, which takes into account nationality within a human rights framework it should be determined as soon as possible after birth. However, the path for banks is long and complicated In relation to politics and human rights; it seems more convenient to keep this focus Arm length.

Therefore, the focus on basic ID fits well with the bank's existing policies important evidence there are many problems with this model. An economic approach to identity can Leading to new forms of public coercion and exploitation of poor population groups and their data Critics of the Aadhaar system point out that it is a private sector. Furthermore, this method conveniently sidesteps the chaotic realities that the national government willfully exclude since.

Apart from legal or de facto requirements, identity platforms are legally neutral. Meanwhile, governments in the global South are saddled with huge debts and spending millions contracting private vendors in public funds to build very easy to work biometric systems A system of exclusion, surveillance and oppression. Banks go out of their way to express this No biometrics are required. By emphasizing their benefits throughout the documentation, the ID4D initiative has helped standardize the widespread use of biometrics in digital ID systems. The World Bank's ID4D programme continues to advance digital identity systems through its three pillars work: thought leadership, network building, and national and regional action published considerable work has been done on the proposed values for these underlying systems.

This contains Develop technical standards for best practice and contribute to the codification of norms The World Bank's Sustainable Development Identification Principles as a Facilitator 'anchor. This expression of the core idea is then used to engage and provide input to other stakeholders the language and resources they need to advance their agenda, strengthening their network of like-minded individuals personal. Furthermore, core parts of the ID4D programme work are often hidden the public perception is that technical assistance is provided to national governments. This includes a formal Diagnostic process, facilitating peer-to-peer knowledge sharing and direct technical support Governments are considering investing in building or updating their ID systems.

Western countries are likely to include basic digital identity systems and controlling immigration and boosting security efforts is an attractive proposition. The UK, France, Germany, Israel, the United States, Australia and other countries are actively developing Invest in research, network building, and often directly in the identification system itself. These countries are interested in strengthening border controls and monitoring their own developments funding and developing new markets for companies including biometric companies. In fact, embassies play a role in contracting out Biometrics Company. Meanwhile, its digital identity system is being touted as a successful government stories like Estonia, India, Peru and Pakistan have positioned themselves as important ID4D partners.

They export "self-developed" ID systems, consultants and expertise directly to others Governments in the Global South. International suppliers have other reasons to be enthusiastic on identification for development as massive investment in digital identity systems means new things opportunity to do business. Suppliers present at major trade shows such as the ID4Africa Annual Conference Information on their products, offering everything from baby biometrics to cloud-based delivery Management information systems and machine learning tools.

The Human Rights Impact of national digital identification systems

Digital identity systems are often built on top of and used in existing paper-based systems Complex environments are overlaid by existing patterns of inequality and social exclusion, so it can It is difficult to separate the meaning of the general formal identification system from the uniqueness.

The impact of digital identification systems

To understand some key human rights issues, we must therefore; two different criteria must be used to evaluate digital ID systems:

- First, we need to build on existing government controls, Formal identification system. Here we have to clearly distinguish between birth systems Registration and civil registration are distinct from other forms of identification procedures. The birth Registration is a basic human right of every child, and so is the civil registration system Indispensable tool of governance and primary means of proving citizenship and other rights.

For example, in several Latin American countries, identification systems have become a means disenfranchised groups have the opportunity to demand recognition in the nation-building process different however, various forms of identification systems also exist, including civil registration systems and national identity cards Vulnerable to exclusion and discrimination in the past. They are also especially appreciated their ability to monitor and control.

- However, the second question is whether the new technology is disruptive or regressive and creates something new implications for human rights. For example, the widespread use of digital biometrics and the reliance on databases and remote verification is entirely due to the new digital system. Additionally, in many cases, digital identity systems are not subject to the same legal and regulatory requirements Models are just like their paper predecessors. For example, Aadhaar does not provide any legal evidence, the identity or residency of the Aadhaar number holder. India's experience has proven both the introduction of biometrics and the decoupling of digital IDs from specific states may lead to new forms of human rights violations, including new patterns of discrimination exclusion and deprivation of political, economic and social rights. Assessing these issues requires careful, in-depth, contextualized research specific to the specific situation country or region.

The introduction of a similar digital identity system is expected to have human rights implications, many political, social and economic lives are already affected by digital identity systems. This Including their right to name, nationality, life, food, education, health, social security and legitimate rights Standard of living, privacy and data protection, freedom of expression and association, freedom of movement, dignity, equal treatment and non-discrimination. Based on this evidence, it would be useful to categorize the impact of digital identity systems on human rights into three distinct areas category:

1. Digital ID systems could increase 'efficiency' of human rights abuses: New launch Digital ID systems often present the same human rights concerns as 'traditional' ID systems Institutions such as discriminatory treatment that lead to the exclusion of social rights and Stateless. One of the functions of formal identification systems is to notice Create hierarchies, exclude "outsider" groups and control population groups. Own clearly, neutral standards can be a means of rendering people stateless. beyond copying Where human rights are an issue, digital identity systems can help governments achieve more By extending existing rights issues to more people and Reduces transparency and engagement by reducing opportunities for remediation and remediation accountability. In addition, centralization of databases and a shift to a "single source" is happening. "Real models" pose significant risks, while the use of new systems leads to the management of a wide range of risks.
2. Digital technologies can introduce novel forms of harm that are difficult to define and address: this includes new forms of exclusion from rights tied to the increased use of biometrics, to digital collection, storage and sharing of identification data, and automated decision-making. This may lead to individuals being shut out or cut off from the system entirely, which combined with the centralization of data and access points can lead to a form of "civil death." It

also includes new types of concerns around the right to privacy, data protection, and cybersecurity, due to the surveillance, control, and manipulation that are afforded to public and private actors under these new digital systems. Finally, digital systems can reverse progress that has already been made on access to rights for marginalized groups.

3. Promised human rights benefits such as inclusiveness and efficiency have not been realized Development of the Weakest: There is an advantage against this damage Basic digital identity systems are currently mostly hypothetical many claims Such a system would enable greater inclusion, especially through financial means Inclusiveness creating efficiencies and savings for governments and promoting economic development, still not supported by a solid evidence base.

A way forward for the human rights movement

A specific digital ID model is being promoted by the World Bank and a wider network of global players this model privileges economic identity, disengages legal status, and directs attention Stay away from civil registration. Contrast with human rights and inclusive development language used in advancing this vision of digital identity, this model threatens a number of fundamental rights, from which from social security rights to privacy rights. The purported benefits remain largely unproven in the absence of serious basic research, cost-benefit and value-for-money analysis and implications evaluate. At the same time, the effects of these systems have been largely studied and disclosed Outsourcing to an already overburdened and underfunded human rights community organisation, lawyers, academics, journalists and other civil society actors against them despite the high stakes, development sometimes seems hopeless.

The World Bank's ID4D initiative published materials emphasizing that the benefits and wisdom of this digital ID vision, ID4D is short for Identification for Development. It relies on the legitimacy of sustainable development goals and human rights provide justification for seemingly inevitable race future. This creates a network of participants who collectively bring about new consensus around digital identities and wrapping them in the language of development goals and rights.

Many groups are competing priorities, resource constraints, and information barriers make it increasingly difficult to respond development of new technologies such as digital identification systems. Digital IDs are all too common the system is bogged down by a lack of transparency.

Furthermore, there are indications that the effectively resisting the false inevitability of a digital future centred on biometric digital data ID system. More and more people have recognized the "dark side" of digital ID, systems and the great work of researchers, activists and practitioners feed us alternative perspectives and reasons for challenge and criticism. National civil society alliances, such as India, Kenya, Jamaica, Uganda and Mauritania all score encouraging victories in defense through their joint research and advocacy, and occasionally opposition to certain digital identity models through litigation. Human rights organisations, other civil society groups, journalists, form a large number of academic experts, learn from each other, and mobilize each other against such a system. This creates increasing opportunities to effect change.

Human rights organisations and other civil society actors can insist on the following:

- Digital identity initiatives that will bring far-reaching changes to governments cannot be rushed without considering their potential impact. Before rolling out new or expanded digital identity systems nationwide, it is important to build an evidence base and take all necessary steps to anticipate and mitigate potential harms. Baseline studies, context-specific studies, cost-benefit analyses, value-for-money analyses, and impact assessments are necessary and should be required at every step.
- Make it public: The design and implementation of new or augmented digital ID systems and any plans in that direction must be discussed thoroughly in democratic forums, including parliaments and public media. Civil society organisation should demand openness with regard to plans, tenders, and the involvement of foreign governments and international organisations.
- We are all stakeholders: The harms associated with digital identity systems are far-reaching and potentially affect all areas of society. Despite efforts to gate the debate and limit it to those with If we have the right technical "knowledge" then we are all legitimate stakeholders. It is important to establish a broad coalition of organisations and individuals who can assess whether digital identity systems can do so whether they contribute to the achievement of human rights objectives.

The rapid rate at which the digital ID agenda is progressing poses a significant problem to those who are slow within the human rights movement and the larger civil society. Many organisations and individuals have they may have been unaware of the recent digital ID evolution in their country or region. It's understandable that many people have not participated in the debates regarding digital ID or other digital forms of identification. Since then, the government has recognized that it is the sector of specialized digital rights organisations.

In addition, the full range and scope of human rights effects that this involves Systems are often misaligned in the early stages and take many years to become fully functional. While catching up on these increases is crucial, the responsibility of proving benefits and avoiding harm must primarily reflect the efforts of powerful actors involved in promoting, designing, building, and maintaining funding digital ID systems at high speed. These actors must have the ability to effectively demonstrate the evidence. The goal is to maximize the benefits and minimize the harm. Instead of hastily advancing, a foundation of evidence and the consideration of the potential effects of these digital ID systems is crucial.

More specifically, this means that human rights advocates and other actors should:

- Request independent studies that focus on the base, digital ID systems are commonly touted as a way to increase participation or enhance power. However, it's impossible to verify these benefits without first understanding the current situation. Even the seemingly obvious question, such as how many individuals are in a country that has access to one or more means of identification may not always be easy to answer. These numbers and figures are important foundations for understanding the future success or failure of new ID systems, and should be considered before implementing any new system.

- Requested analyses of cost-benefit and value for money: Despite the common claims about the supposed economic and other benefits of digital ID systems, there are few serious analyses of this type. However, if an accurate and democratic evaluation of the wisdom of implementing a new or augmented digital ID system is to be conducted, the various costs and benefits must be considered in advance. Many factors need to be considered, including the initial cost of designing and building a digital and biometric system, the associated costs of maintaining the system's registration capacity and necessary human resources, and the financial viability of the system. Crucially, the monetary and non-monetary costs associated with individuals, especially those who are marginalized and discriminated against in a particular context, must be taken into account, as must the effects of reliance on electricity and the internet connectivity, as well as the cost of fees within the system. Other critical factors to consider include the reliance on foreign companies, cybersecurity concerns, and the opportunity costs associated with various factors.
- Ask for impact evaluations: Even with the most dedicated preparation preceding this, all actors should expect the impending. As a result, it would be beneficial to demand that any new digital ID system's roll-out be undertaken with caution and respect. Because of the possible harm, digital ID

systems can cause, it's important to consider these effects carefully and then evaluate the results. Other than the costs and benefits associated with these new systems, the impact of these on human rights should be a significant component of these evaluations, these institutions should have sufficient resources and independence to do this.

Discussions and decisions should occur in public forums like parliaments and media outlets, the latter of which should have watchdogs and other mechanisms to monitor decisions. The government should provide the greatest degree of transparency regarding plans and implementation, all pertinent information should be accessible in the public domain, and the government and international organisation should be transparent about the evidence and baselines that support the adoption of the digital ID system at hand. Organisations can participate in these processes and promote greater transparency by asking questions and conducting their own research as often as possible.

One of the main issues associated with the digital ID agenda has been the difficulties experienced by human rights organisations and the civil society in comprehending what is happening and what is being promoted and implemented.

The following questions may provide some guidance for organisations on the types of questions to ask about existing and proposed digital ID projects.

Key questions to ask about national digital ID projects:

- At what point in the digital system of identification is the government? Many countries have already discussed potential collaborations with actors like the World Bank for several years. Others have begun to procure technology, create legal and regulatory structures, and implement systems. Other individuals have employed a digital ID system for years, whether or not the Bank participated in it. Understanding the frequency of critical design decisions regarding system implementation and the 'state we are in' should be the first step for anyone interested in taking part.

- Why is the government advocating for a new or enhanced digital ID system? It's crucial that the civil society can comprehend and evaluate the justification for the chosen digital ID model, and receive an accurate description of what causes the development of digital ID, from national security to social inclusion. Additionally, they must have the ability to question the benefits and costs of the proposed project. This involves understanding the supposed benefits as well as the theory that supports them. Therefore, civil society organisations must consider: what is the issue? What evidence is there of the issue? Can the proposed system or existing system address this issue? Might other non-intuitive solutions, or solutions based on other forms of ID have a lower cost? Therefore, civil society organisations should promote the government conducting a cost-benefit analysis of concrete proposals, and should also involve relevant experts in the assessment of benefits and costs. Through their own investigations, civil society organisations can additionally expose other goals, which may cause the distribution or expansion of the digital ID system.
- What legal regulations or safeguards already exist? What legal mechanisms are in place to protect rights, including

privacy and access to justice, for example? Is the formal law that defines the national digital ID system effective? Does it necessitate the requirement to conduct human rights impact studies? This includes legislation and regulations pertaining to cybersecurity, the exclusion of social rights, as well as the legal definitions of the relationship between different ID systems. How does the new digital ID system interact with the existing civil registration systems, or other individuals' already possessed ID's, including private methods of proving identity?

- What impact will the new funding for a digital ID system have on the current civil registration and vital statistics system? How will this impact the funding of and organization of birth registration? Will the responsibilities of government be regarding registration shift, for example, from social departments in charge of civil registries to new organisations or agencies? Will new organisations like National ID's or digital government agencies be created? How will these new entities differ from other government parts? What initiatives are underway to sustain the funding of the national ID, and ensure that the Databases are current and complete? Will funding be shifted away from the civil registration system due to the introduction of a new digital ID system?
- Which international organisations are involved in funding, design, and construction of the digital ID system? Do international or national banks for development participate in the process? What is the degree to which they participate? How will they help the model's operational and financial viability? It's crucial to demand copies of their

recommendations to governments, proposed funding, and other forms of participation. Additionally, where other international organisation dedicated to development are involved, what is their participation? Civil society organisations can invite these organisations to participate in public hearings and other forms of government that are open to the public, these forms of government allow organisations to expose foreign actors to the public's gaze.

- Who private foundations and donors are involved in funding or contributing to the national ID system? What is their current role and what do they propose? How will they help the model's operational and financial viability? What inspires them? Can these individuals be encouraged to utilize their influence in beneficial ways? Civil society organisations can explore ways to educate donors about the true nature of digital ID.

Governments often rely on the same small group of international organisations, foundations, and technology providers for guidance, inspiration, and financial assistance in the design or enhancement of their national digital ID systems. The World Bank is only one part of a larger network of international organisations, foundations, corporations, and other entities that are involved in the development of digital ID systems. However, it has attempted to occupy the centre of this network, and is currently serving a significant role as an internationally recognized advisor to governments, in advocating for the legitimacy of digital identification processes, in creating the global network, and in funding digital ID projects. However, there are still more participants who have a significant role in these discussions.

Hundreds of organisations dedicated to civil society and numerous experts around the world have already devoted their efforts to issues associated with digital ID systems, as well as to questions surrounding the registration of births, and the capacity for statistical analysis. Other individuals work with communities directly affected by digital ID systems and regarding the biometric and digital components of these systems.

There is already a diverse range of knowledge and expertise that helps civil society organisations understand the complex landscape of digital ID systems. However, it's crucial to recognize that everyone has a role in systems of identification, digital or otherwise, which are crucial to recognizing individuals and exercising their human rights. Through a collective, cooperative effort to assemble the existing knowledge of digital ID systems and their effects in different contexts, we intended to explain the global consequences of these systems. Directing attention to the expansive, often ignored, network of actors involved in the digital ID agenda, we intend to assist future efforts to advocate for digital ID.

Digital IDs unlike traditional physical IDs are stored on a computer or mobile device and can only be retrieved using special identification methods such as fingerprints. Digital IDs are used by governments around the world for a variety of functions, such as voting, accessing government services, and online identity verification. The use of biometric information is another form of digital identification that is gaining acceptance. Biometric information includes unique physical characteristics that can be used to identify an individual, such as fingerprints, facial features, and iris patterns. Governments use biometrics to generate digital identities that can be used for a variety of purposes, including access to financial services, international travel and online identity verification. Because biometric information is difficult to forge or steal, it is considered a more secure form of identity.

Blockchain-based digital identities are also being developed to provide secure and decentralized identities. These identities are stored on the blockchain, a decentralized ledger that securely and transparently records transactions. Blockchain-based digital identities can be used for a variety of purposes, including voting, accessing financial services and verifying online identities. They are considered a more transparent and secure form of identity because they cannot be easily forged or tampered with. Several countries have also introduced digital passports as a form of digital identity. Digital passports store an individual's biometrics, personal information and travel history in a digital format that can be accessed from a mobile device.

Digital passports can be used for cross-border travel and are considered a safer and more efficient form of identity verification than traditional paper passports. Finally, some governments use social media data as a form of digital identity. Social media data includes information about individuals' online activities, including their posts, likes and comments. Governments use this data to create digital identities that can be used for a variety of purposes, including access to government services and online identity verification. However, the use of social media data for digital identity has been criticized because it raises privacy and data protection concerns. In summary, governments around the world are developing and implementing various digital identity systems to provide secure, efficient, and convenient ways to verify an individual's identity. While these systems offer many potential benefits, they also raise concerns about privacy, data protection, and the potential for discrimination or exclusion. It is therefore imperative to carefully consider the impact of these systems and ensure that they are designed and implemented in a way that protects the rights and freedoms of individuals.

The complexity of digital ID and biometric implementation

Biometric and digital ID technologies involve the collection and storage of sensitive personal information, such as facial features, fingerprints, and other biometric data. This information can be used to identify individuals with great accuracy, making it an attractive option for governments, businesses, and other organisations looking to improve security and streamline operations. However, it also raises privacy and data protection concerns, as this information could be misused or leaked if it falls into the wrong hands. One of the biggest challenges in implementing digital ID and biometrics is ensuring that the right safeguards are in place to protect individual privacy and prevent data misuse. They are personal. This includes implementing strong privacy and security measures such as encryption, firewalls and access controls to prevent unauthorized access and data leakage. It also requires clear and transparent policies regarding the collection, use and disclosure of personal data, and mechanisms by which individuals can exercise their rights to access, rectify and delete their data.

Another challenge is ensuring that digital identities and biometrics are available and inclusive for all members of society. This is especially important for refugees, the homeless, and vulnerable and marginalized populations without access to mobile devices or reliable internet. Ensuring that these people are not excluded from essential services and opportunities requires innovative solutions that take their unique needs and circumstances into account. The use of digital IDs and biometric data also raises important questions about civil liberties and options for state surveillance and control. These technologies have the potential to be used to monitor and track an individual's activities, movements and associations, and may be used to restrict an individual's rights and freedoms. It is therefore important to ensure that the use of these technologies complies with the appropriate legal and regulatory framework, with appropriate controls and countermeasures in place to prevent misuse, exploitation and abuse. In addition to these challenges, the implementation of digital ID and biometrics is complicated by the need for international standards and interoperability.

This is necessary to ensure that these technologies can be used across different jurisdictions and across different organisations without unnecessary barriers or inconsistencies. Overall, the implementation of biometric and digital ID technologies requires a careful balance of potential benefits and associated risks and challenges. This requires a collaborative, multi-stakeholder approach that takes into account the needs and concerns of all stakeholders and ensures that appropriate safeguards are in place to protect human rights, privacy and civil liberties.

Your digital identity could be used against you in the event of a major reset

When a radical solution to a hyper generalization problem pushes the boundaries of privacy and individual liberty, that solution poses a greater threat to everyone than the problem it was designed to solve. As part of a "Great Reset" agenda to reshape the global economy, the World Economic Forum wants everyone on the planet to join a technological identity number. While digital identities promise to improve livelihoods for millions, authoritarian governments can also describe and Control the behaviour of citizens in the social trust system.

The idea behind digital identities is pretty simple. All the data you collect from every online interaction you have with the private and public sectors makes up your digital identity. In terms of personal convenience, having a digital identity that consolidates everything in one place can be a godsend because you can use your digital identity anywhere to purchase a variety of goods and services, and all your data is in the zone protected on the blockchain. There is a big difference between identity and identification. It's one thing to give ID to someone without official documents.

It's a completely different story when this concept is expanded to link every person on the planet with digital identities that record their actions. Identity is everything that makes you unique, and it is who you are that really matters to the World Economic Forum. When someone crosses the line, every social media interaction you engage in, every penny they trade, and every action you take can work against you The Chinese Communist Party has adopted the IoB concept to create an Orwellian surveillance state, linking digital identities of the "network" with a social credit system.

The “Unscrupulous Debtors” contact tracing app alerts citizens when they come within 500 meters of the DNA phenotypes of debtors who have been placed in refugee camps for more than 1 million Uyghurs. The CCP or Chinese Communist Party is a living example; there are many scary ways digital identities can be exploited. In some post-lockdown areas, you can't leave your home, you can't buy or sell certain products, you can't open your own shop. Authorities will be alerted if you breach a curfew, enter a restricted area or open a shop after being asked to close as recommended by the World Economic Forum.

If the unelected elites get their way, you will own nothing

You have no privacy; your every move is tied to your digital identity as you wait to be approved or denied your benefits under a "citizen-centred welfare state". And you better do as you're told, because every action is tied to your digital identity. Cross the line, and every social media interaction you engage in, every penny you trade, and every move you make can be used against you.

A vaccine passport is a form of digital identity

A vaccination card is essentially a digital identity; the COVID passport credentials power an authoritarian social credit system through a digital identity system. As a result, citizens around the world have come to rebel against their governments' arbitrary orders and restrictions that have nothing to do with public health, but only with control. Vaccination cards set the stage for widespread adoption of digital identities. Digital identity systems are now laying the foundation for a global social credit system linked to financial institutions.

Central Bank Digital Currencies as a Means to Introduce Digital ID

A CBDC linked to a digital ID could allow governments and businesses to set limits on what you can buy with your own money, including expiration dates on which you can spend it. Once digital IDs and CBDCs achieve a level of acceptance and mainstream acceptance, the option to return to physical media can be quietly and recklessly removed.

Convenient as a point of sale for digital IDs

Simplicity is a key selling point for public and commercial organisations using digital identities. Elites rely on convenience and the illusion of choice to gain mass acceptance and trust. The efforts of public and private organisations to justify the adoption of digital identities seem endless. Globalists and unelected governments say digital identity systems are needed to mitigate the effects of COVID-19 and cyberattacks, and advocate for the elimination of anonymous transactions in favor of a fully traceable, interoperable, and fully programmable CBDC. Finally, digital identity systems could enable governments and corporations to manipulate, coerce, or change people's behavior, while removing individual authority, freedom, choice, dominance, and anonymity under social credit systems.

The WTO and WEF promote "global digital identification of persons and items"

For those who manage data and set the rules for digital identification, "We the People" are in danger of becoming a "product as a service". The World Economic Forum and the World Trade

Organization want to give physical objects, digital goods, and biological people digital identities, just as the so-called Fourth Industrial Revolution merges our physical, digital, and biological identities. Any container, pallet or cargo can now be fitted with sensors, transmitters or RFID tags, allowing companies to track its performance and usage as well as its location in the supply chain. A digital identity framework is being used to support vaccination passports, it is being transformed into a Metaverse passport perspective, which is a new kind of digital passport.

The G20 called for the use of data on social and economic activity to implement the 2030 Agenda. Digital identity architects have played a key role in its development as the digital identity framework has become increasingly complex with the addition of biometrics. Central banks are now considering the idea of letting anyone use a digital ID to access a central bank digital currency. However, installing digital ID and biometric systems is not without its difficulties.

One concern is that individuals' digital identities could be used against them during a major reset, vaccination cards are a digital identity promoted by governments and the globalist elite. The Bank of America cyberattack was used to facilitate the launch of digital currencies that were evaluated as a way to establish digital identity schemes. A universal digital identification of people and things is the goal of the World Economic Forum and the World Trade Organization.

Chapter 6

Digital Currency and Blockchain

In our current monetary system, money is dualistic. The system is based on currency, which is notes and coins, as well as commercial bank reserves that are held in commercial banks at the central bank. Banknotes and coins are the most common forms of currency that we all recognize from childhood, however, the practice of using currency to pay for things varies from nation to nation. The second form of currency that is commonly understood is deposits, which are maintained in bank accounts. Despite having no physical counterpart, we can utilize the money to pay someone else's account by asking our bank to debit our own and simultaneously credit someone else's account. In macroeconomics, cash and other monetary assets, as well as the deposits of households and non-bank corporations in commercial banks are considered the money supply.

The monetary funds derived from central banks that possess physical properties, such as banknotes and coins, or those derived from central banks that lack physical properties, such as commercial banks' reserve accounts at the central banks, are the foundation of our current monetary system. Most importantly, private deposits in the commercial banking system are augmented with a second layer. This is also illustrated by the fact that deposits with banks, like the majority of other financial obligations of banks, are ultimately composed of central bank money. For instance, a deposit is the right of a financial institution to issue notes and coins as needed. Since commercial banks cannot create money on their own, deposits in accounts are the actual debt of the bank.

Despite the fact that we keep most of our money in deposits today, the monetary term that describes the system is still relevant. Since deposits lack any physical appearance, transfers of funds simply represent alterations to the system of books or banking. While these old bankbooks were previously large collections, today they are primarily electronic data banks that are stored on servers. Additionally, the majority of our payments for deposits are conducted via payment cards, online banking websites or smartphone apps. Banks and payment card providers utilize the most recent safe data transmission methods and complex cryptography to guarantee the legitimacy of every transaction.

Modern currency deposits are actually virtual currency deposits, the rise of digital currency like Bitcoin and its associated blockchain technology has been of great interest. These advances offer the potential for significant effects on the financial system and the larger economy. As such, authorities and central banks worldwide have been surveying the evolution of digital currency and evaluating its effects in recent years. A frequently asked question is whether or not central banks should create their own digital currency that is publicly accessible. This chapter will focus on the primary reasons that central banks will issue digital currency, it also explores the consequences of this move. Ideas of how to eliminate cash or eliminate large notes are typically derived from two causes. First, increasing the expense of criminal activity, black market or money laundering, and tax evasion. The second is to increase the number of monetary policy options available. The first rationale is debatable.

It's unclear to what degree the practice of cash promotes criminal activity and illegal behavior, and the balancing of the benefits of avoiding crime with the costs of cash-based restrictions on legitimate activity. Also, dropping cash may have a negative impact on financial knowledge. Additionally, preventing cash from being used for purchases diminishes privacy, which has both social and private value. In any case, removing cash is difficult because in a free economy, any government alterations to the monetary system are confined by the necessity of maintaining the government's currency as a popular choice for users.

The second motivating factor has its flaws; the starting point is the realization that banks that issue deposits cannot decrease the rate of these deposits below zero without putting large amounts of cash on the line, which would lead to a destabilizing effect for the institution. The lower limit of the deposit rate is effective, which means that the lower limit of the central bank's policy rate is also effective, unless the central bank is prepared to reduce the financial institution's spread of credit, which will have a negative impact on the economy. The profitability of the bank and the potential for financial stability. Destroying or taxing cash, or regulating its usage, would reduce savers' external options and lead to more extreme monetary policy actions.

Central banks are preparing to replace all cash with a digital currency CBDC Central Bank Digital Currency. This gives them complete control over humans, as they can monitor every transaction, limit your spending, and even wipe your funds at will. It's advertised as something that makes our lives easier. The truth is, this would mean total tyranny of all humanity. Our digital funds will be linked to our digital IDs, which will be linked to our vaccination status and social credit score. A CBDC uses a digital ledger to record every transaction you will make. Similar to Bitcoin, except the ledger is permanently linked to your digital ID. Anyone with access can see everything you buy and sell, every bill and tax you may or may not have paid, and use whatever they find against you at any point in the future. Laws and rules are constantly changing, rarely for the benefit of ordinary people.

Money with a maturity date means you will never be able to save and create financial stability for yourself or your family. No savings to buy a car, a house, or even a rainy day. No cash in birthday cards, no money from the homeless or street musicians, who might even call it "carbon credits." Everyone lives from scratch and cannot plan anything. There is another word for people who cannot own property slaves. Programmable money means banks can also limit where and when you spend your money and what you can buy. It can easily be linked to a personal carbon allowance, so you only have a certain amount of fuel or meat each month. With the push of a button, your wallet can be frozen or restricted within the distance you can use it from the registered address. Lockdowns are easy to implement and don't require a single policeman.

While CBDC proponents have proposed many potential benefits, these do not hold up to scrutiny. In short, these proponents fail to draw any meaningful distinction between CBDCs and the digital dollars that exist today. But CBDCs are not just a story of government waste or cronyism. While CBDCs offer no particular benefit to the public, they do pose serious risks to financial privacy and economic freedom. From expanding financial regulation to destabilizing the financial system, a CBDC could impose enormous costs on a country's citizens. In short, there is no reason for the federal government to issue a CBDC when the costs are so high and the benefits so small. Congress should ensure that the federal government does not issue a CBDC.

What Is a Central Bank Digital Currency?

A CBDC is a digital national currency. In the case of the US, the CBDC would be a digital form of the US dollar. Like paper money, a CBDC would be a liability of the Federal Reserve. But unlike paper money, a CBDC would not offer the privacy or finality that cash does. Indeed, it is this digital responsibility, a digital link between citizens and the central bank that differentiates CBDCs from the digital dollars already in use by millions of Americans. In the private sector, Americans now routinely use various forms of digital dollars. They send digital payments through credit cards, debit cards, prepaid cards, and various mobile apps. In fact, not only payments have gone digital. Almost every financial institution offers everything from savings accounts to mortgages through mobile apps. Therefore, there should be no misunderstanding that the US dollar is already widely used in digital form.

Furthermore, the current system works so well that few people take the time to consider whether the digital dollar they use is Visa debt or Bank of America debt. While there's no reason for consumers to think twice, when using a prepaid card, the balance on the card is the responsibility of the private company that issued the card. When consumers deposit money into their bank accounts, the deposits in those accounts are also liabilities of the bank. Effectively, this means that the bank owes the customer the funds deposited into the account. When a consumer transfers money from a bank account, the bank is responsible for the transfer. However, in the case of a CBDC, the digital dollar would be the responsibility of the central bank itself. This means that the government, in the case of the United States, the Federal Reserve, is directly responsible for holding, transferring or otherwise distributing these funds to the so-called transfer owners.

This function provides a direct link between citizens and the central bank, in contrast to the existing system in the United States where private financial institutions provide banking services to retail customers. Perhaps because this departure is so radical, some CBDC proponents are touting an "intermediary CBDC" in which private financial institutions will "serve" accounts. Under this arrangement, account balances remain on the Fed's balance sheet. However, private financial institutions will provide all necessary retail banking services. Most policy implications are the same for brokered and non-brokered retail CBDCs.

Many CBDC proponents claim that a CBDC would improve financial inclusion by providing new sources of financial services to the underbanked and unbanked population of the United States. However, like proponents of other misguided proposals to improve financial inclusion, CBDC proponents ignore the innovation already happening in the private sector and the aspirations of the unbanked. Proponents also claim that maintaining the US dollar's status as the world's reserve currency is a potential benefit of a CBDC, but even that falls short. The dollar's prestigious status is due to the strength of the US economy and its legal protections for individuals compared with most other countries, not to the specific technology that enables electronic funds transfers.

If US Congress wants to strengthen the role of the dollar, it should focus on improving these underlying causes, not the recent enthusiasm for central banks. For example, establishing stronger financial privacy protections, removing government barriers to faster payments, and requiring greater transparency in monetary governance could benefit the dollar's international status. None of these steps involves issuing a CBDC. One of the reasons cryptocurrencies have become so popular is that they have become the main option for citizens of many other countries whose payment systems are weak and unreliable compared to the US system. The US should also go back to the gold standard, which is a monetary system in which paper money is freely convertible into a fixed amount of gold giving the US Dollar more value.

There is no risk that the US dollar will lose its status because the Fed does not have a CBDC, especially when countries adopting CBDCs offer few economic and legal protections that belong to the US system. For example, China's eCNY which is a Blockchain version of a CBDC, is unlikely to attract global demand given the Chinese government's long history of violating property rights, financial privacy, and other human rights. Likewise, Nigeria's CBDC eNaira is unlikely to attract global demand given Nigeria's history of volatile inflation and volatility. Finally, the Sand Dollar, the Bahamian CBDC, is unlikely to attract global demand, as the Bahamian Dollar is not sufficiently linked and the country's economy is not strong enough to be an international reserve currency. Therefore, while each country faces unique challenges in achieving world reserve status, CBDCs cannot solve anything.

Monetary and Fiscal Policy

Proponents also argue that a CBDC could improve the conduct of monetary and fiscal policy. It is claimed that a CBDC will provide an opportunity to improve the economy on an individual level, open the door to charging negative interest rates, and remove credit and liquidity risks in the market. All three statements are false. Suggestions that CBDCs could allow policymakers and regulators to fine-tune the economy are both encouraging and worrisome. Whether it's the 1970s or the 2010s, it's no secret that the Fed has struggled to achieve its policy goals. Indeed, given the Fed's poor track record in managing price levels and the business cycle, it makes perfect sense that the Fed has worsened overall stability.

Still, some believe that by tracking Americans' financial activity, the Fed could eventually steer the economy in a positive direction. For example, some CBDC proponents believe that the use of negative interest rates is a missing tool in the Fed's arsenal. While CBDCs may be new, negative interest rate proposals are not. Proponents have long called for a ban on the use of cash to implement negative interest rates, and using a CBDC to implement such a policy would likely require a ban on all alternatives, such as cash, cryptocurrencies, and foreign currencies. Finally, many proponents argue that CBDCs can attract more customers than the private sector because they offer an alternative without credit risk and liquidity risk, but this distinction is misleading. The technology behind CBDC offers no such advantages. These zero-risk features are entirely attributable to government guarantees, which can of course be added to any private sector electronic payment option. In other words, if Congress wanted to ensure that certain payments were processed without liquidity or credit risk, they could do so without a CBDC.

The general public has a right to privacy, but financial privacy has been limited for decades. Laws aimed at fighting terrorism, preventing money laundering and collecting taxes have largely empowered the government to conduct uncontrolled surveillance of financial information. Still, with few safeguards left, a CBDC may be doomed as it would provide the federal government with complete transparency of every financial transaction, creating a direct link between the government and the financial activities of every citizen. There is currently a buffer between government financial activity and the public because this activity is spread across all the different commercial banking and payment services that people use. The government may not need to obtain a warrant to access much of the financial information held by these various companies, but tracking them down and working with them provides a buffer, albeit a small one. However, if the government offers a CBDC, it will no longer have this buffer. All financial data will be just a keystroke away.

The threats that CBDCs may pose to liberty are closely related to the threats to privacy. With so much data and consumers so connected to the central bank, a CBDC would offer governments countless opportunities to control citizens' financial transactions. Such controls can include prohibiting and restricting purchases, restricting purchases or freezing and confiscating funds. The programmability of a CBDC could lead to people being banned from buying certain goods or limiting the amount they can buy. For example, advocates quipped that parents could set aside money for their children's lunches, on the condition that it wasn't spent on chocolates. It is important to consider the expansion possibilities of this choice.

Just as parents try to control their children, policymakers may try to curb alcohol consumption by restricting late-night purchases of alcoholic beverages or prohibiting those with alcohol-related offenses from purchasing alcohol. In the event of a government-mandated lockdown during a pandemic, a CBDC could be programmed to only transact with "essential" businesses or alert authorities when citizens incur travel expenses. The possibilities for CBDC programmability are virtually endless. In all cases, even the best intentions are only a few steps away from leading to serious abuses of power. Aside from the basic programmability, CBDCs provide for social and political control, one of their most common features is the ability to pay positive and negative interest rates to both disincentives and incentivize purchases. This means that government providers of CBDC can easily deposit funds directly into citizens' accounts and withdraw them with equal ease.

Ultimately, the imposition of this negative interest rate penalty depends on consumers having no other means of payment, which is one of the reasons why governments have adopted CBDCs to ban cryptocurrencies. Nevertheless, not only third-party competitors must be eliminated. Central banks cannot impose negative interest rates on CBDC if people can continue to use cash (paper money). In fact, getting rid of cash is exactly what many CBDC proponents are calling for. More broadly, governments have long recognized that freezing a person's financial resources is one of the most effective ways to exclude a person from society.

A changing economy

Financial markets in Latin America have long been plagued by uncertainty due to a lack of trust in traditional financial institutions. This lack of trust is exacerbated by rampant corruption and instability in the region's social spheres, especially economically. In response, people turned to digital assets to protect their wealth. Rising inflation is also helping to accelerate the development of remittance infrastructure and asset tokenization in Latin America. This is an important step forward for the Latin American digital asset space. Authorities have set standards for companies to operate, created rules, and are working on a central bank digital currency. Several companies in Latin America have been investigating ways to use blockchain technology and digital assets to solve the myriad problems currently facing countries in the region.

The CCP will use digital money in China to use a social credit system to penalize or force citizens

The Chinese Communist Party is developing a central bank digital currency as part of a larger social credit system that would allow the government to monitor, control and direct public behavior. With this system, the Chinese government hopes to process and support more than 300,000 financial transactions per second, while feeding the information into the Communist Party's real-time social credit system. Under China's emerging social credit system, citizens are given credit scores based on their online and offline behavior. It's a system that rewards "good" behavior, like spending time with older people, and punishes "bad" behavior, like protesting the government or spending too much time playing video games. When "trust" is broken in one area, restrictions are imposed everywhere, meaning citizens who violate even minor rules are barred from traveling, eating out, renting a home or even buying insurance.

According to Chinese state media, this has happened to more than 30 million citizens. The central bank's new digital currency and electronic payment system, driven by big data, will give the CCP another authoritarian weapon to monitor and monitor citizens' behavior. CNAS assesses the integration of real-time financial data into the CCP's technology-driven governance plan, "This push for digital currency is an important step in strengthening the party's digital authoritarianism." Even if it doesn't use blockchain technology, the system could "facilitate the CCP's economic surveillance and government intervention in the lives of Chinese citizens." The CCP is not only interested in financial transactions; it also combines this financial data with data from the Internet of Things, smart homes, CCTV footage, and the dynamics of mobile phone usage, including citizen geolocation and browsing history.

How the blockchain is being used as a tool for social change

Blockchain technology is increasingly being used as a tool for social change. Every industry including legal services, healthcare, agriculture and many others is poised to be transformed by it. The cryptocurrency Bitcoin is the most famous application of this technology. However, the underlying technology is finding a way to be more widely used. The basis of virtual currencies such as Bitcoin and Ethereum is blockchain technology. This is open ledger technology. Every transaction is efficiently and permanently recorded. Any programme can run on it. Blockchain is also transformative for many social groups, these groups are already harnessing the socially transformative potential of this technology. They think it's a great way to attract customers and build trust.

Preserving safe documents

The technology provides the ability to securely store transaction records. This is especially useful for tracking medical records. Healthcare professionals face obstacles when it comes to storing patient records. With blockchain technology, all barriers are removed, at least to a manageable extent. The technology exposes security features that traditional sharing techniques cannot provide. In this case, most privacy concerns are satisfactorily resolved. This reduces the number of intermediaries that are allowed to process the input. Because it maintains ongoing transactions, it reduces the chance of error and promotes a high level of trust.

Sharing information

This technology could facilitate the decentralization of Wikipedia's knowledge base by allowing editors to participate in the network as stakeholders. Relying on the power of technology makes user contributors more accountable. Editors can earn tokens based on their significant contributions. These donations represent a stake in the platform. It encourages correctness of posts; as wrong changes can cause contributors to lose tokens. There is no central server where data is stored. As a result, the data will remain even after the central organization collapses. Then the government cannot filter the data. The government cannot stop anyone from donating.

Anti-corruption

This technology has the potential to be an effective weapon against corruption. It protects data and guarantees the integrity of records thereby ensuring authenticity. It eliminates the chance of data management fraud. Blockchain supports a decentralized method of recording transactions. Immutable chains are enabled by recording each transaction in chronological order. Public blockchains help law enforcement detect illegal activity because the technology leaves a digital trail when identifying suspects.

Student Records

Blockchain technology is not only beneficial to the healthcare industry; it can also be used to benefit the education sector. A student's academic record, including grades, certificates, courses, and other information can be stored in a blockchain application. Data from different educational institutions can be centrally managed. It makes it easy to capture digital records and use them as references. It allows access to data while protecting the integrity and privacy of teachers and students. Potential employers can obtain job references from reliable sources.

Potential use

Most people are still not clear about the concept of blockchain. It's like going back to the 1990s with the idea of the Internet. People don't fully understand the concept yet. For now, some see this as an abstract idea. Still hard to understand. However, it is becoming increasingly clear how it can be used to address many of the challenges facing organisation of all kinds. In the financial sector, it guarantees payment and guarantees instant payment without going through a financial institution. When signing a smart contract, both parties no longer need to hire an intermediary or a third party to perform the contract. Additionally, this technology allows advertisers to engage with users interested in their services and reward them with tokens. The technology doesn't depend on a single controller; it uses code and algorithms to control everything.

It is revolutionizing industries and changing the way people work. The potential of this technology is enormous. It makes a significant contribution to society as it reinforces public integrity. It can also restore citizens' trust in public institutions. It also adds another layer of security to every transaction and record. However, it is important to remember that this is not a panacea. The technology is still under development. It may take years for pilot projects to become widely available. Again, it's important to understand what its worth to society and how to deal with people's inflated expectations. It's not a magic wand. This requires stronger institutions. But technology can effectively amplify them.

In summary

Digital currency and blockchain technology are two of the most popular buzzwords in the world of finance and technology. A digital currency is a form of money that is completely digital and exists only in electronic form. Unlike traditional fiat currencies, digital currencies are not backed by a central authority such as a government or financial institution. Instead, they rely on cryptography and complex algorithms to ensure their security and authenticity. One of the most popular and well-known digital currencies is Bitcoin, which was created in 2009 by an anonymous individual or group named Satoshi Nakamoto. Bitcoin and other digital currencies are decentralized, meaning they are not controlled by a central authority, and their transactions are recorded on a public ledger called a blockchain. A blockchain is a distributed ledger maintained by a network of computers around the world. Each block in the chain contains a set of transactions, and once a block has been added to the chain, it cannot be modified or deleted. This makes the blockchain very secure and transparent, as anyone can view the entire transaction history of a digital currency. Blockchain technology has many potential applications beyond digital currencies. It can be used to build decentralized applications, or dApps, which are software applications that run on a blockchain network. These applications can be used for a variety of purposes such as decentralized finance, voting systems, supply chain management, and more. One of the main benefits of blockchain technology is its security.

Because blockchain is decentralized, there is no single point of failure that hackers can target. Transactions on the blockchain are also secured using cryptography, which makes them nearly impossible to crack or alter. Another benefit of blockchain technology is its transparency. Because all transactions are recorded on the public ledger, anyone can see them and verify their authenticity. This makes blockchain ideal for use in industries where transparency is critical, such as finance, healthcare and supply chain management. Despite the many benefits of blockchain technology, there are still some challenges that need to be addressed. One of the biggest challenges is scalability, as current blockchain infrastructure is not yet capable of handling the transaction volumes of traditional financial systems.

There are also concerns about the environmental impact of digital currencies, as the mining process for some cryptocurrencies requires a lot of energy. However, many in the industry are working to address these challenges and make digital currencies and blockchain technology more accessible and sustainable. Collectively, digital currencies and blockchain technology have the potential to revolutionize the way we think about money, transactions, and trust. As technology advances, we can expect more and more innovative use cases and applications to emerge that have the potential to transform industries and create new opportunities for individuals and organisations.

Chapter 7

Transhumanism

A worldwide movement known as transhumanism claims that human augmentation is achieved through the implantation of DNA and technological implants. However, transhumanists gloss over two issues: the use of technological implants as a weapon against ordinary people, and their process of carrying out risky projects that could be a cover for illegal human experimentation around the world. Technological implants such as brain nanobots could lead to a loss of mind control, which would result in the wearer being manipulated by others, losing autonomy and being permanently monitored via the internet of the brain. If their memories are wiped, their identities are lost. People implanted with technology are thus vulnerable to constant surveillance, mental manipulation, and loss of identity, making them human slaves working for multinational corporations and political and economic superpowers. Objective analysis shows that transhumanism is simply an intellectual hoax leading to digital fascism, a society ruled by multimillionaire elite citizens with technological implants who will become digital slaves in the service of an oligarchy.

Transhumanism is considered an expansion of humanism, from which it derives in part. Humanists advocate that individuals are significant, that people are significant. We may not be flawless, but we can improve things by promoting rational thinking, freedom, tolerance, democracy, and caring towards others. Transhumanists concur, but also suggest what we can become. Just as we employ rational methods to enhance the human condition and the outside world, so we can also employ this method to enhance ourselves, the human body. Other approaches to humanism, such as cultural and educational development, are not limited by the traditional definition of humanism. Other technological advances can ultimately allow us to be more than others consider "human". It's not the specifics of our biological makeup or our human form that define our value, but rather the goals we seek and the life we lead. For transhumanists, progress is the ability of more people to alter their behavior, their lives, and the way they interact with others in accordance with their deepest beliefs. Transhumanists value autonomy highly: the capacity and right of individuals to choose and plan their own lives. Of course, some individuals are reluctant to utilize technology to enhance themselves due to a number of reasons.

Transhumanists attempt to create a world in which autonomous individuals can choose to remain as they are or choose to improve, in doing so, they will honor their chosen options. With the rapid rate of technological advancement and scientific understanding, we're now entering a whole new era in human history. In the near future, we may have to deal with the reality of true artificial intelligence. New methods of thinking that combine both artificial intelligence and interface technology are being created. Molecular nanotechnology has the potential to provide a lot of resources to everyone and allow them to control the biochemical processes in their bodies, this would enable them to avoid disease and prevent aging.

Technologies like brain-paced interfaces and neuropharmacology can augment human intelligence, increase emotional well-being, enhance the capacity to commit to long-term projects or loved ones, and even increase the range and complexity of possible emotions. On the dark side of the spectrum, transhumanists acknowledge that some of these upcoming technologies have the potential to cause significant harm to human life; even the survival of our species may be in danger. Attempting to comprehend danger and attempting to prevent a catastrophe is integral to the transhumanist philosophy. Transhumanism is now commonly accepted in culture as more and more scientists, trained in the scientific method, and social entrepreneurs begin to take seriously the many possible applications it provides.

According to the Global Transhumanist Association, transhumanism is an ideology that supports the use of technology to help people transcend our biological limitations and improve the human condition. This trend has been described as beneficial to society, an advancement of humanity, a continuation of humanity, or even a "revolution." "According to transhumanists, the post-human species will emerge as a result of adding technological implants and injecting DNA into humans. Humans will stop biological evolution and begin technological evolution. The integration of nanotechnology, biotechnology, information technology and knowledge the so-called "four new technologies" of science are applied to human beings, which is the essence of transhumanism.

Transcendentalists believe in using technology to improve people beyond their current physical and mental limitations. Their goal is to create healthier, faster, fitter, smarter people. For them, it is time for humans to take responsibility for their own evolution, which will give rise to a new species of Homo sapiens. These "transsexuals" or "offspring" can live up to 500 years. Transcendentalists envision a world in which the human body is obsolete and must be replaced by "better vessels" such as robots. They see the body as a flawed and obsolete object that needs technology to eliminate and replace. The aim is to implant silicon, steel and microchips into the human body. The changes brought about by implant technology in the body will make transgender people truly asexual, and their gender will be changed beyond recognition.

The metahuman movement was started by Julian Huxley in 1957, and in the 1980s the first officially recognized self-proclaimed metahumans met at the University of California, which became a major centre of metaphysical thought. Transcendentalists have founded political parties in several countries, and even some churches have adopted their ideology. The transhumanist movement has the support of renowned scientists and professors from the most important universities in the world. The campaign uses clichés such as "innovation," "evolution," "excellence," and "digital citizenship," as well as various ad campaigns promoting intellectuals and entrepreneurs.

Despite their unrealistic, almost illusory claims, metaphysical theories are promoted and advocated at the highest official levels in the world and are promoted by the media on a daily basis. They encountered little resistance from scientists or intellectuals, and they even preached their unrealistic assumptions. As a result, prestigious institutions such as universities and governments in the United States and Europe consider sublime as part of their programmes. In Peru, for example, the Registry of the National Identity System promotes transhumanism through campaigns such as "digital transformation" and public lectures at universities.

The main deception leading to digital enslavement is transhumanism

In general, transhumanists hide the negative effects of technological implants, they do not use the brain implants they promote, economic powerhouses and multinational technology companies provide billions of dollars in support and funding for transhumanism, and Experimentation using illegal human implants is considered the truly secretive method of these transhumanist projects. All of this suggests that transhumanism is a hoax designed to make us dig deeper.

Microchip

An integrated circuit, commonly called a microchip, is a tiny electronic device consisting of a complex network of electronic components on a thin silicon wafer. These components include, but are not limited to, transistors, resistors, capacitors, and diodes. The introduction of microchips, which enabled the development of smaller, faster and more powerful electronic devices, is considered one of the most important technological advances of modern times. Microchips are used in everything from smartphones and computers to medical devices and cars.

Transistors act as switches that regulate the movement of electricity and are the basic building blocks of microchips. Engineers can build complex circuits that can perform a variety of tasks by fusing millions of transistors onto a single chip. The technique, known as photolithography, uses light and chemicals to etch small patterns onto silicon wafers, enabling large numbers of transistors to be placed on a single chip. Depending on their intended function, microchips come in different sizes and shapes. Some microchips are much larger and are used in industrial applications, while others are much smaller and are used in medical devices such as pacemakers.

Is a future where everyone is chipped becoming closer?

While many people shy away from invasive techniques, the risks associated with using microchip implants may not be as great as many people think. Many people fear genetic engineering or implants that work like robots with the human brain. It can be argued that this fear is either real or imagined. Whichever perspective you choose, there's no denying that these situations have significant societal implications. Therefore, all steps in this direction should be painfully incremental. However, isn't the risk/benefit analysis much easier when it comes to implants that don't directly interact with the body?

What are implants with microchips?

Microchip implants are passive devices implanted subcutaneously in the body that contain integrated circuits or RFID tags encapsulated in quartz glass. Thanks to the housing of the gadget, the integrated circuit does not come into contact with the body, so there is no physical reaction and the device is protected. Smartphones have plenty of storage space. Passive implants have long been used in cattle and livestock. Because of these benefits, acceptance for use in animals continues to increase. RFID implants can aid in traceability, breed and pedigree identification, and theft prevention by providing information about animal ownership.

Does this also apply to people with microchips?

Although not a very new technology, it does not yet have wide acceptance and acceptance among the people. More than 20 years ago, the first human implant was performed in 1998. Since then, only a small group of biohackers have used it. While these devices are not commercially available for human use on a large scale, that may soon change. If people think a chip with their banking details is destructive, there is nothing more invasive to us personally than delivering a drug to our body.

What would happen if a third party had access to the RFID records and could alter or disrupt the system?

Before microchips become commonplace among the general public, too many questions remain unanswered, including whether governments and corporations have the right to plant alien objects in your body that collect information and Provide medication. But don't worry everything will be perfect. "After all, you have nothing to hide." Overall, human microchipping has been proposed for more than 20 years. A tiny electronic device called an implant is placed under the skin, usually on the hand. These chips are used for a variety of purposes, including location tracking and health monitoring, as well as conducting transactions and opening doors.

Despite the potential benefits of implanting microchips, adoption of the technology has been slow. While entrepreneurs are working to make the technology more widely available, privacy and security concerns remain. Critics argue that implanting microchips into the human body could lead to the establishment of a surveillance state and the possibility of data collection. The possibility of collecting personal data is one of the concerns. A generation of microchip biohackers has exacerbated the problem. Manufacturers and implanters of these chips will have access to personal information and will be able to monitor people's whereabouts, purchases and health information.

The potential for misuse of this data is a serious concern. The security of the chip is also an issue. In the event of a hack, the chip's data could be compromised, which could lead to identity theft and other security breaches. Additionally, the potential for third parties, such as governments or employers, to access and manipulate data on chips raises civil liberties concerns. The concept of implanting microchips into the human body also raises ethical questions.

What are the long-term health effects of having a foreign body implanted?

Beyond microchip implants, DNA barcoding has the potential to change the way we monitor people and things. DNA barcoding can be used to track everything throughout the supply chain, including people, with precise data recovery. Although DNA barcoding technology is available, there are ethical and legal issues with its use. While the concept of microchips in humans has been around for some time, the development of the technology has been gradual. While entrepreneurs are working to make the technology more widely available, privacy and security concerns remain. In addition, it is important to address the potential for misuse of technology and ethical issues. Although. DNA barcoding has the potential to change the way we monitor people and things. Their use also raises ethical and legal questions.

Face Identification

Face recognition technology has been widely used for many years, and it is a way for computers to recognize and recognize human faces. The face recognition system is a technology that can compare the face in a digital photo or video frame with a face database. This technique locates and measures facial features from images and is widely used to verify individuals with identity verification services. The development of such systems began with computer applications in the 1960s. Since its inception, the face recognition system has been widely used in technological fields such as robots and mobile phones.

Computerized facial recognition systems fall into the biometric category because they recognize physiological traits unique to humans. Although facial recognition systems are not as accurate as iris and fingerprint recognition as a form of biometric identification, they are widely used due to their contactless nature. In addition, facial recognition systems are used for automatic image indexing, video surveillance, and advanced human-computer interaction. Governments and businesses around the world are now using facial recognition technology. However, the effects have varied, and several methods have been abandoned due to failure. The facial recognition system has also drawn criticism, with claims that the system invades residents' privacy, routinely misidentifies individuals, supports racial and gender stereotypes, and promotes gender norms. Additionally, deepfakes and other synthetic media raise questions about the safety of traditional media. The allegations have led to facial recognition technology being banned in some US communities.

Not all of the facial recognition systems are dedicated to recognizing individuals based on a database of matches. Some systems are intended to assess the general population's trends or sentiment, these systems are designed to scan large crowds of people with facial recognition. Contrasting with the traditional video surveillance system, which has been a staple of law enforcement for decades, facial recognition cameras are more intrusive. You can identify important facial features in order to create a comprehensive biometric profile of a person that does not require their permission. Surveillance cameras that use facial recognition are typically portable and sneaky. For example, the security forces of Malaysia have partnered with the Chinese technology company Yitu to provide officials with cameras that recognize faces. This will facilitate a comparison of the images captured by live cameras with those in a central database that is quickly done. Citizens are now refusing to use facial recognition technology. Hong Kong protestors, for instance, covered their faces and utilized destabilising apps on their smartphones to deny access to law enforcement. They also altered the way police officers are identified by taking pictures of them without their ID's and using computerized facial recognition to identify them online.

Methods for recognizing faces

While face recognition in computers is a difficult pattern recognition problem, humans can recognize faces with little effort. Based on its two-dimensional representation, face recognition algorithms try to recognize a human face, which is three-dimensional and changes its appearance based on lighting and facial expressions. The four stages in which the face detection algorithm completes this calculation are: First use face detection to separate the face from the background of the image. Then, in a subsequent stage, the segmented face images are aligned to take into account the face pose, image size, and captured features such as illumination and grayscale. The goal of the alignment process is to allow the facial feature extraction method (third stage) to precisely locate facial features. Details such as eyes, nose, and mouth are then identified and quantified to represent the face in the image. The fourth step is to compare the determined face feature vector with the face database.

Traditional

Certain facial recognition algorithms extract landmarks or features from images of a person's face to identify facial features. For example, the relative position, size or shape of the eyes, nose, cheekbones and jaw could be checked algorithmically. The next step is to use these features to find similar features in other photos. Some techniques normalize mugshot collections, compress facial data, and store only facial recognition-related data from the images. The face data is then matched to the probe image. One of the first effective systems used a collection of key facial features as input to a template-matching algorithm to create a compressed facial representation. Geometric algorithms, which focus on distinguishing features, and photometric algorithms, which use statistics to reduce an image to a set of values and compare those values to a template to remove variation, are the two main approaches used in detection algorithms. Others divide these algorithms into two broad categories: feature-based models and ensemble models. The latter decomposes the face into parts and analyzes each part based on its spatial position with other features, while the former tries to recognize the face as a whole.

Detecting people from a distance

Facial hallucination is used to enhance low-resolution photos of faces, making it easier to identify people at a distance. The faces in CCTV images are often very small. Facial hallucination techniques are performed using nearest neighbour distribution indices or example-based pixel replacement machine learning before photos are passed to a facial recognition system, which may also include demographic and age-related facial features. Face hallucination methods help high-resolution face recognition algorithms work better and can be used to circumvent some of the shortcomings of high-resolution algorithms. Face hallucination techniques are also used to pre-process images with face masking. After removing camouflage (such as sunglasses), the image is processed using a face hallucination algorithm. These face hallucination algorithms must be trained on images of concealed and incomparable faces. Facial hallucination algorithms must accurately map the full state of the face to fill in areas exposed by uncovering, which may not be possible given the fleeting facial emotions captured in low-resolution images.

Recognition of 3D

3D face recognition methods use 3D sensors to collect data about the shape of the face. This information is then used to identify the contours of the eye sockets, nose and chin, and other identifiable features of the facial surface. An advantage of 3D facial recognition is that, unlike previous methods, it is not affected by changes in lighting. It can also recognize faces from different angles, including side views. Three-dimensional face data points greatly improve the accuracy of face recognition. The development of sophisticated sensors that project structured light onto the face has enabled 3D facial recognition research. Researchers at Techniker used metric geometry tools to treat expressions as isometrics, since 3D matching techniques are sensitive to expressions. In a new technique, three tracking cameras are used to capture 3D images of faces. Three cameras were used, one facing the subject from the front, one from the side, and one tilted. These cameras work together to track the patient's face in real time, allowing for facial recognition and identification.

Infrared cameras

Using a thermal camera is another way to collect data for facial recognition; in this process, the camera only recognizes the outline of the head, ignoring any subject accessories such as glasses, hats or makeup. Additionally, the thermal camera can record images of faces even when it's dark outside at night without using a flash and exposing the camera's position. Still, there aren't many facial recognition databases available. Scientists at the US Army Research Laboratory have developed a method that compares images of faces captured by thermal imaging cameras with those in a database of facial images captured by conventional cameras. This technique of creating a single image by examining many facial regions and features is known as transfer synthesis because it combines facial recognition from two different imaging modalities.

It consists of an optimization problem that projects latent projections back into image space, and a nonlinear regression model that transforms a given thermal image into the corresponding visible face image. All studies agree that the method is successful because it integrates local and global information, such as facial features Facial recognition technology works by examining a person's facial features and comparing them to a database of recognized faces. The system uses various algorithms to identify and identify facial features such as eyes, nose and mouth. It then looks for matches by comparing these attributes to a database of recognized faces. The use cases for facial recognition technology are numerous.

Retailers could use it to identify repeat customers and deliver targeted marketing, while law enforcement agencies could use it to track down suspects in criminal investigations. Facial recognition technology, despite its potential benefits, has raised concerns about civil rights and privacy. According to critics, the technology is error-prone and could be used to track people without their knowledge or consent. As a result, the need for further regulation and control of facial recognition technology has increased.

The Technology's Consequences

While fraud prevention and authentication are generally positive things, there are several implications for the technology and its deployment. Technology itself is advanced, but it must be used and treated responsibly.

Is it fair if an insurance company denies people health insurance based on the predictive power of facial recognition technology?

Asia and the West have different views on these issues. China is a country that doesn't take privacy as seriously as it does here. The technology is being deployed haphazardly across the country without any regulations or personal protections.

China Merchants Bank has launched ATMs that only require facial recognition so customers can withdraw cash. The technology is used by several of Asia's largest employers as a site access control system and identity mechanism. It offers many amenities. China is a country with an authoritarian government that demonstrates how technology can be used to monitor and control its citizens rather than empower them. Chinese citizens are using facial recognition technology to bolster their country's controversial social credit system. Governments are working hard to influence the behaviour of citizens, and this is possible through existing technologies. Protesters in Hong Kong recently took to the streets wearing ski goggles and surgical masks to evade identification by the Chinese government. This provides new evidence of how facial recognition is being used against citizens.

Should the government be open about using face recognition technology?

Rather than trying to use facial recognition technology behind the public's back and then saying it's for the greater good, transparency from the government about its use of the technology would be much more welcome. The US Department of Homeland Security already uses facial recognition software at biometric entry points at US airports, despite growing criticism of the technology.

Creating a database of photos of millions of people could lead to civil rights violations, according to opponents of facial recognition. Future citizens may face awkward situations when such databases are shared with other agencies, including police authorities or private companies. Keeping an eye on your back can become a habit. The Sociable has previously noted that facial recognition software has also been deemed too biased and imperfect to adequately identify real criminals or suspicious people. Nonetheless, there is no doubt that there is an urgent need to improve airport security measures. Relying solely on human vision is no longer recommended.

Chapter 8

Smart Cities

A smart city is an urban environment with ubiquitous surveillance and data collection technologies that monitor and record everyone's most intimate personal information. The goal is to know everything about each person: what you eat and drink, where you go, what you buy, who you meet, what you think, how you feel, your opinions, your habits, your health and vaccination status etc. The rationale for this all-encompassing surveillance is that it is considered essential to saving "the planet" from climate change. They also guarantee that your life will become more connected, safe and healthy. Every aspect of smart city life is monitored through various data collection technologies, traffic and many other SMART technologies.

Together, they form a ubiquitous surveillance network, constantly collecting all information on every detail of people's lives. These external SMART devices sync with devices that connect directly to the body, such as the SMART watches on our wrists or the SMART phones in our hands. These devices are able to gather information about what is going on inside our bodies and even have the ability to change how the body functions. In the near future, they will be implanted in the human body to collect more data and have a greater impact on what happens in our bodies. Personal information collected by billions of devices around the world forms a network known as the "internet of the body".

The internet of bodies is essentially the same, except that it will collect all the personal and most private information about everyone. This data will be in the hands of companies and governments around the world. To ensure everyone is within range of data collection, travel in smart cities will be limited to 15-20 minutes from home. The concept coined the term "15-minute city" and has been touted as the best idea ever to save the planet from climate change. Everything you need is within a short distance, so no one has to go beyond this area. Long distances require a permit.

The discussion of the Smart City and academic research on the impact of urban digitization has traditionally been led by computer science, data science, and engineering. Concocting scientifically-based methods, technical procedures and technological innovations, the primary research goals are the mathematical modeling, simulation, data mining, remote sensing, machine learning and large unstructured information and data. Discussions regarding the political, economic, cultural, and social effects of urban digitization have since taken place, the development of theories, papers, and predictions were all significant in the 1990s. The business and technology industries have had a significant impact on the development of the smart city agenda, and have made it a reality that more and more public functions and services are delegated to private entities. Smart cities have been demonstrated to promote the privatization and corporatization of public spaces, resources and services.

Large corporations participate in the smart city movement and are often responsible for the nature, timing and adoption of technology in city government. Capital and economic forces promote the smart city processes that are driven by neoliberal, colonial, and imperialist ideologies. Cities, services, and public goods are now being utilized by companies to increase their revenue, and are increasingly being considered markets for companies to attempt to earn the highest profits. Urban policies that promote private actors have led to the emergence of a small, wealthy elite that represents the common good instead of the general interest. Cities that have a capitalist political philosophy have a tendency to increase inequality and spatial division, the ICT and smart city initiatives are undoubtedly involved in the present and future of urban life. The combined effects of urbanization and digitalization since the mid-20th century have led to a transformation of cities into smart cities in recent years. This has had a significant effect on cities and their residents. Smart cities revolutionize the way cities are governed, governed, and operated, altering the way cities provide services, gather information, and respond to threats. Additionally, they open up new methods for citizen participation, mobility, and interaction with the community.

Engineers and computer scientists from the ICT industry, as well as municipalities, collaborate to develop new solutions to urban problems, and utilize smart cities as a means to improve the quality of life, increase economic growth, and improve the efficiency of urban management and service delivery. Initially, however, little concern was devoted to the social and political consequences of these technological advances. With few exceptions, only recently have social scientists dedicated their research to the study of how digitalization affects city residents and urban life, this has led to the subject of smart cities being studied as a form of research. They advocate for a variety of social criticisms of smart cities, which augment the discourse of STEM-based smart cities.

Social scientists acknowledge the dangers of overemphasizing technological solutions and the difficulty of having to wear technical glasses. They demonstrate the association between digitalization of cities and top-down political government, and acknowledge the dangers of technology companies' increasing influence and control over urban politics. They advocate for the privacy, surveillance, and security concerns of city residents, as well as the dangers of smart city policies that increase the division in urban communities and exclude marginalized groups. Many of the largest urban problems are not technological in nature, but social, political and economic, thus they need to be addressed with political solutions rather than technological ones.

A common theme in this social critique is the lack of humanity in the discourse about smart cities, and the depiction of smart cities as mechanical, controlled systems that are unimpeded by the complexity of human beings. Many advocates of the smart city movement have responded to criticism and attempted to make smart city initiatives more centric, but many of these attempts have been unsuccessful and have merely altered the discussion, not the underlying principles that are involved in the development and implementation of technology in smart cities.

Smart cities and technological innovations can increase the social divide, enhance corporate power, and lead to top-down control without taking into account all of the residents. A smart city should first cater to people and base its actions on the value of each individual's specific abilities and requirements. A human rights framework can explain how to create smart cities that focus on people, utilize technology in a way that promotes and preserves human rights, has an equal approach to solving problems, and targets marginalized individuals and groups. The Right to the City and Human Rights the Urban Movement promotes the transformation of language and formal processes of political decision-making in favor of democracy, equity, justice, inclusion, and reciprocity. The degree to which a city is considered "smart" is influenced by both technological advancement and social metrics, as well as the degree of tolerance. This is the place where human rights have a particularly significant role: they can re-conceptualize priorities, establish standards for the evaluation and measurement of urban policies, resist non-conventional practices, and observe the implementation of equitable and inclusive approaches.

A human rights approach to smart city management necessitates addressing complex social, political and economic issues in order to take precedence over issues that can be accomplished through technology. It mandates that human rights, including civil, political, economic, social and cultural rights, be implemented and incorporated into local policy processes. It necessitates the involvement of multiple stakeholders in all parts of the smart city movement, including urban planning and policy development, this is especially true of those who lack the institutions necessary to shape the urban environment. Ultimately, and most importantly, a human rights approach must transcend simple statements and empty words, it must actually empower and release city residents. City governments and technology companies must find ways to encourage citizens to participate in the creation and enhancement of smart cities for human rights. This implies surrendering some influence and participating in the decision-making process from the bottom up. Smart cities that prioritize citizens over other stakeholders are likely to be leading in the smart city movement, they will explore new markets and innovations, and they will be the first to advocate for the rights of all humans.

George Orwell's 1984

George Orwell wrote his major work 1984 during a turbulent time in world history, his novel 1984 was published on June 1949, World War II ended three earlier. The last truly global conflict made Western citizens aware of the pitfalls that people in totalitarian regimes can run into. Nazi Germany controlled its citizens through constant fear of being watched by agents of the Gestapo, the secret police, convincing ordinary Germans to betray their families and friends in opposition to the Nazi state. The defeat of Nazism in Europe prompted Europeans like Orwell to reassess the world they lived in because Orwell, a former supporter of Leninist communism, recognized that Stalin's Soviet Union was in many ways very similar to Hitler's Germany resemblance. This way of thinking undoubtedly drove Orwell in 1984. Orwell wrote an extreme picture of totalitarianism in 1984. The novel is set in futuristic Oceania, a country made up of the former England. The inhabitants of Oceania lived under the rule of the Ingsoc political party. Ingsoc is the newspeak word for the political ideology of the governing party of Oceania and is short for English Socialism.

Ingsoc, under the omnipresent leader "Big Brother", controls every aspect of its members' existence. Party members lived under the constant surveillance of the telescreen. These acted as two-way devices, allowing Ingsoc to spy on its officers and spread propaganda. Party members risk severe punishment for minor indiscretions, such as not showing the correct facial expression. The 1984 storyline follows Winston Smith's troubles: a party member who privately refuses to take the party line. When Orwell died in 1950, his vision of a society living under the complete surveillance of state technology would have struck most people as ludicrous. The novel 1984 can be read as an allegory of our current living conditions. I will argue that other elements of Orwell's vision are already evident in our contemporary world. New media technologies redefine global boundaries. My main argument is that new media technologies impair free thinking in a manner similar to that Ingsoc uses to automate its followers. I will critically assess how new media technologies have changed the human mind.

Winston Smith lived in Oceania, a country governed by unwritten codes of conduct. Ingsoc will punish any member who violates these guidelines. The problem is that the rules of Ingsoc are not only unclear, but also very unstable. This means that the citizens of Oceania may one day be wiped out for behavior that was perfectly acceptable the day before. The same is true for us as we live in a global society mediated by new media technologies. Winston Smith lives in a world where surveillance is the norm. Citizens of Oceania watch over their neighbors and families at all times. If a person is suspected of speaking negatively about Big Brother, they can be vaporized. In Oceania, children are exposed to a culture of surveillance from an early age. Ingsoc led his own youth movement called The Spies. The technical training young spies receive allows them to spot potential enemies of the state. One of Smith's associates, a man named Parsons, was told he had been arrested over his 8-year-old daughter. Every member of Ingsoc lived under the constant surveillance of the telescreen.

A telescreen is most effectively described as a two-way system of CCTV that through the telescreen, the leaders of Ingsoc surveyed the lower class citizens of Oceania to make sure they were behaving according to protocol. If a citizen of Oceania were to wear the wrong facial expression, they would be reprimanded by their leader for the offense. All members of the Ingsoc Party were monitored by the 'Thought Police' on a continuous basis. Orwell borrowed from previous societies of observation in order to model Oceania. The society's structure paralleled that of Nazi Germany and the Stalinist USSR, both of which were totalitarian. Both countries had a supervisory system that was similar to the Nazi's and the Stalin's. The purpose of this system was to eliminate or reprimand individuals who did not conform to the Nazi's or Stalin's ideology. In both countries, the same was true of Oceania as described in 1984. A sculpture of Oliver Cromwell was situated in the town square of London, this sculpture functioned as the symbolic representation of Oceania's capital. Cromwell had previously been a controversial political figure in the UK during the 16th century. The Cromwell's and their followers overthrew the monarchy and took control of England, Scotland, and Ireland. The regime of Cromwell's followers was extremely brutal in its treatment of any perceived threat. In Orwell's Oceania, as in Britain's Cromwellian, unconventional freethinkers were not permitted.

Ingsoc utilized the creation of propaganda as a viable method of controlling the populace of Oceania. Orwell based his promotional state on Nazi Germany and the USSR. It's my contention that various organisation in our real world utilize the same promotional tactics. I would maintain that the US government has been primarily guilty of this behavior. The Ministry of Truth, which Winston Smith held the position of chief of government, took over the initiative for the creation and distribution of propaganda across the Oceania region. Ingsoc employed their network or television show to disseminate false information to their constituents.

The US government has attempted to influence our thoughts and emotions through the use of various technologies. Some of the agencies of the US government have utilized new media in their efforts to influence our thoughts and emotions. In the following paragraphs. Both the Ingsoc and the US government have employed digital tactics to explain and promote global warfare in Oceania, it was one of the three countries in the world in 1984; the other two were Eastasia and Eurasia. The Ministry of Truth reported that Oceania had been consistently involved in conflict with either the Eurasian or Eastasian nations in succession since the countries were formed. Smith observed that the country alliance or opposition moved around on a regular basis. Through the media, Ingsoc disseminated a narrative that altered the historical record, this narrative portrayed Oceania as always having a conflict with the same adversary. The US has utilized digital media in this manner to promote its digital war effort among the American population in order to gain support for it. Social networking and online gaming are both new media forms that the US government has employed to promote its military agenda.

The Ingsoc's mantra of 'he who controls the past controls the future' is apparent in the many political propaganda methods employed by governments in our modern world. It can be hypothesized that regimes, including the US, utilize new media to undermine narratives that are in opposition to their own by subverting the perspectives of others, the US guarantees that its narrative of the truth is the most popular. The effects of one central authority that possesses a monopoly on the truth are not necessarily positive. As a result, it's not uncommon to make a direct comparison between Ingsoc and the US government. Both the Ingsoc and the US governments have attempted to monopolize the truth, utilizing digital technology for communication. It's apparent that the US government would generally support Ingsoc's philosophy that 'we don't simply destroy our enemies; we alter them' in an era of digital technology and new media, it seems that the truth is malleable.

We live in a society that is characterized by the frequent use of new media in everyday interactions. It's become more apparent that the communication abilities that these technologies provide us have a significant negative impact on our personal liberty. Our recreations of Big Brother take various forms, some of which are the agencies of different world governments. The US government employs similar surveillance and propaganda tactics to those of Ingsoc. Recent world events have demonstrated that the US government is readily committed to censoring individuals who have the freedom to think and speak freely. Individuals who publicly disavow the US government's narrative are subjected to severe punishment. The increasing prevalence of netspeak via new media is detrimental to the critical thinking and communication abilities of our youth population. To replicate Orwell's methods, until we become aware of the ways new media is being employed against us, we will never take rebellion. Our awareness is contingent on our desire to rebel.

Difference in Classes

Oceania is divided into three classes: Inner Party, Outer Party and Proletarians. Big brother looks at all courses. The core party members account for 2% of the population of Oceania, but they are the real masters of Oceania. They formulate policies, make decisions and rule the entire city. The Inner Party, commonly known as the Party, represents the upper class, and they enjoy different rights than others, such as temporarily turning off screens, living in luxurious homes, enjoying good food and wine, having personal servants, and the ability to use helicopters and cars.

It's not easy for anyone to hack into their building for no reason the class differences in Oceania are used to give more power and control to the party and its allies, and the people under the party rule are physically and mentally oppressed. Moreover, the class diversity in Oceania has helped the party empower government leaders over ordinary citizens, leading to its repression. Forced into poverty, the people of Oceania were reduced to the underclass and oppressed by the party leadership.

Additionally, children have a crucial role in the regime of totalitarianism. Spies that are told to follow things by the Party, even if they are told to keep an eye on their parents' information, these spies believe that it is necessary to be devoted and loyal to the regime. Mr. Parson was the victim of his daughter, but he was happy about her.

What is more apparent in 1984 is the fact that the working class is overly influenced by Party promotional campaigns, these campaigns attempt to explain to them the purpose of the Party and the Othello situation in their conflict with the opposition. As a result, the proles believe that it's an exceptional thing that benefits them. While the actual purpose of propaganda was to enhance the Party's control over the Oceania population, The Party instead sought power for its own sake.

We have no concern for the happiness of others; we have only a desire for power, absolute power. As a result, this propaganda facilitates the distinction between classes because it promotes a means of controlling and dominating people, Orwell intended to demonstrate the ways that are employed in a totalitarian government that converts people into objects in its possession.

The culture depicted in Oceania throughout the story of 1984 is similar to the typical society today. Increased government activity is causing a more tyrannical regime to emerge, this has significant consequences. Our benefactor system of preserving individual liberty has been supplanted by an overbearing government that is now spying on its constituents in order to demonstrate its power. Less civil liberty is associated with a stronger, more tyrannical government, the example of which is Oceania, and this is now common in many countries. Today, people live in countries that are similar to Oceania, and there are class differences.

Digital Propaganda

Authoritarian regimes are increasingly utilizing digital operations of influence as a means of censorship and manipulation, they flood the information space with false or misconstrued stories that aim to drown out independent voices and knowledge. Their aspirations may be as simple as attempting to misinterpret facts surrounding specific incidents, or as complex as attempting to wreck institutions and democratic cohesion through the use of social, political, and economic resources. Increasingly, evidence is presented that authoritarian networks are increasing in size, spreading across species, and learning from one another.

Several terms have been employed to describe various information space activities; these include hybrid warfare, psychological warfare, active measures, fake news, disinformation, propaganda, and coordination of inauthentic behavior, information/ influence operations. While not replaceable, they all describe a variety of associated malign behavior that is intended to mislead or deceives, in the global information space. The term "digital influence operations" will be employed to describe the common types of digital activity utilized by authoritarian regimes internationally to influence, regulate, and maintain the integrity of the information space for strategic purposes. While these initiatives take place in the digital realm and are intrinsically connected, they do not involve robots, or programs that are automated online. Because of the widespread nature of bot activity during global democratic elections in recent years, the perception that inauthentic collective action is the entirety of these efforts can lead to misguided responses.

Actually, authoritarian digital enforcement efforts utilize all components of the information space, including through the ownership of digital media outlets and technology platforms, business and advertising, and traditional methods of censorship. In the case of the PRC, this includes a variety of initiatives intended to influence the architecture, standards, and governing bodies of the global information space in a way that benefits the Chinese Communist Party and promotes a related restriction on free speech, this is distinguishable from the effort of democratic governments.

Individual countries have different strategic goals, and they have dedicated different methods to achieving these goals, but they have also attempted to take the best practices from each other and amplify them when it is beneficial to them. Many authoritarian nations have a shared goal of not only destroying their own images on the international stage, but also instilling distrust in democracy and the rule of law in general. Discrediting democracy as a form of government is a goal that authoritarian regimes pursue, the expense of doing so via the methods described here has decreased significantly in recent years. Additionally, many authoritarian countries have a reliance on information and narratives to maintain their security, these countries believe that controlling these aspects is crucial to their foreign policy. The following section focuses on some of the key countries' digital manipulation methods and procedures, but it is not intended to be complete.

Different varieties of polities have attempted multiple social mechanisms for regulating propaganda. In a perfect democracy, everyone would have the opportunity to promote their cause and counter promote the cause of others through non-violent methods. The democratic ideal supposes that, if multiple different types of propagandists are permitted to compete without restraint and publicly, the most beneficial ideas will ultimately win. This outcome would require that the majority of the population be educated, intelligent, and devoted to the public, they would also need to not be overly confused or upset by the communication's excess. A democratic system also requires that large amounts of dependable and pertinent information will be cost-effectively disseminated by public-spirited, news-gathering agencies that are well-endowed. The degree to which any existing society in the country is actually consistent with this model is admittedly a question of open debate. That the social system of the world is not obvious.

The regime would attempted to monopolize all opportunities to engage in propaganda, it will also stop at nothing in order to destroy any form of counterpropaganda. How long and how completely this policy is implemented is dependent on the amount of power the regime has, the degree to which they are thorough in their police work, and, most importantly, the distribution of secular higher education. Secular higher education typically promotes skepticism about claims that are based on lack of evidence or are overly dogmatic; if the education is of a humanist or Universalist nature, an authoritarian regime that is ignorant of the educated will not be able to satisfy them for a long time. If the educated individuals participate in discrete counterpropaganda, they may ultimately alter the regime.

Conclusion

As this book draws to a close, it is abundantly evident that the world is about to undergo a significant transition toward a digital future. Governments all around the globe are setting an example by putting various programmes into place that aim to reduce the use of cash in society, introduce digital currencies, and establish digital IDs, among other things. We will emphasize some of the concerns that we need to be aware of as we go toward this digital future in this conclusion, as well as a summary of the major themes discussed throughout the book.

It is plainly clear as this book comes to a close that the world is set to go through a big change toward a digital future. Governments all around the world are setting an example by implementing a number of initiatives that, among other things, aim to minimize the use of cash in society, introduce digital currencies, and establish digital IDs. In this conclusion, along with a recap of the key ideas covered throughout the book, we will highlight some of the problems that we need to be aware of as we go into this digital future.

Digital Authoritarianism: The rise and global spread of digital authoritarianism in China is changing the balance of power between democracies and authoritarian states. China provides the world with a blueprint for a digital totalitarian state. All democracies and civil society actors must unite to resist the global spread of Chinese digital authoritarianism and defend and preserve freedom and dignity in the 21st century.

AI: Artificial intelligence arouses both fear and enthusiasm. Many see advances in artificial intelligence as an opportunity, while also worrying that AI could end humanity. If human values can be carefully articulated and embedded in AI systems, socially unacceptable outcomes can be avoided. So the questions that needs to be asked is does artificial intelligence represent an opportunity or a threat? Will machines take all jobs or create more jobs than they destroy? Opinions vary, what is certain is AI will continue to transform the world of work, and workers will need to engage in lifelong learning, develop skills, and change jobs more frequently than in the past.

The social credit system, which is now being implemented in China, is another important issue discussed in the book. Data analytics will be used by the system to evaluate residents' actions and determine whether to reward or penalize them. Although the system may have certain advantages, such as promoting moral conduct and enhancing social trust, it also presents serious privacy problems and the possibility of government misuse.

Digital ID and biometrics: We have also looked at the development of digital IDs and the use of biometrics to increase security and decrease fraud. Digital IDs are being created to replace conventional paper-based IDs, and biometrics like fingerprints and face recognition are being utilized more often for authentication. We have, however, already covered the dangers of using biometric data, including the possibility of privacy loss and data breaches.

Digital Currency and Blockchain: The book also discusses the emergence of cryptocurrencies like Bitcoin and other types of digital currency, as well as how blockchain technology is being applied to make transactions safer and more open. We have spoken about how digital money might lead to greater financial inclusion and cheaper transactions. The concerns connected to digital currencies, such as volatility, cybersecurity, and the possibility of unlawful activity, have been underlined as well.

Transhumanism: The concept that technology may be utilized to increase human capacities and longevity has finally been discussed in the book. We've spoken about how technological developments like artificial intelligence and nanotechnology may usher in a world without humans. But we've also looked at the ethical ramifications of transhumanism, such as how it may lead to a new kind of inequity between those who can afford the improvements and those who can't.

It's crucial that we keep these concerns in mind as we transition to a digital future and weigh the advantages of technology against any possible drawbacks. To guarantee that these measures are carried out in a way that protects individual freedoms and rights while fostering a more just and sustainable society, we must cooperate. To develop best practices, standards, and laws that serve the public interest, governments, business executives, and civil society groups will need to work together.

We are going toward a more digital future, as seen by the development of the cashless society, the social credit system, digital ID and biometrics, digital money, blockchain, and transhumanism.

Although these technologies have the ability to completely transform our society, we also need to be aware of the dangers and difficulties they provide. Together, we can build an inclusive, egalitarian, and sustainable digital future.

Made in the USA
Monee, IL
05 November 2026

67811467R00085